AF837922

Electric Vehicles

This book offers a thorough exploration of electric vehicle (EV) technology, spanning from its historical roots to its current state and prospects. It provides a comprehensive overview of the various aspects surrounding electric vehicles, aiming to educate readers on the technology, environmental impact, market trends, challenges, and potential solutions within the realm of electric transportation, and also:

- Provides an in-depth exploration of EV technology, covering topics such as types of EVs, propulsion systems, battery technology, and charging infrastructure.
- Includes case studies and practical examples, offering readers valuable insights into the application of EV technology in the real world.
- Discusses the future outlook of EVs, including emerging technologies, ongoing research, and potential advancements.
- Explains the technical aspects of EVs such as battery technology, power electronics, and regenerative braking systems.
- Showcases factors like carbon emissions, energy sources, and lifecycle assessments, allowing readers to understand the broader environmental footprint of electric cars compared to traditional vehicles.

This book is primarily written for graduate students and academic researchers in electrical engineering, electronics engineering, communication engineering, automobile engineering, and energy engineering.

Electric Vehicles

Advanced Technologies, Case Studies, and Applications

First Edition

Dolly Thankachan
S. Sangeetha
Shweta B. Thomas

CRC Press
Taylor & Francis Group
Boca Raton London New York

CRC Press is an imprint of the
Taylor & Francis Group, an **informa** business

MATLAB® are trademarks of The MathWorks, Inc. and are used with permission. The MathWorks does not warrant the accuracy of the text or exercises in this book. This book's use or discussion of MATLAB® or Simulink® software or related products does not constitute endorsement or sponsorship by The MathWorks of a particular pedagogical approach or particular use of the MATLAB® and Simulink® software.

First edition published 2026
by CRC Press
2385 NW Executive Center Drive, Suite 320, Boca Raton FL 33431

and by CRC Press
4 Park Square, Milton Park, Abingdon, Oxon, OX14 4RN

CRC Press is an imprint of Taylor & Francis Group, LLC

© 2026 Dolly Thankachan, S. Sangeetha and Shweta B. Thomas

First edition published by CRC Press 2026

Reasonable efforts have been made to publish reliable data and information, but the author and publisher cannot assume responsibility for the validity of all materials or the consequences of their use. The authors and publishers have attempted to trace the copyright holders of all material reproduced in this publication and apologize to copyright holders if permission to publish in this form has not been obtained. If any copyright material has not been acknowledged please write and let us know so we may rectify in any future reprint.

Except as permitted under U.S. Copyright Law, no part of this book may be reprinted, reproduced, transmitted, or utilized in any form by any electronic, mechanical, or other means, now known or hereafter invented, including photocopying, microfilming, and recording, or in any information storage or retrieval system, without written permission from the publishers.

For permission to photocopy or use material electronically from this work, access www.copyright.com or contact the Copyright Clearance Center, Inc. (CCC), 222 Rosewood Drive, Danvers, MA 01923, 978-750-8400. For works that are not available on CCC please contact mpkbookspermissions@tandf.co.uk

Trademark notice: Product or corporate names may be trademarks or registered trademarks and are used only for identification and explanation without intent to infringe.

ISBN: 978-1-041-00083-9 (hbk)
ISBN: 978-1-041-00085-3 (pbk)
ISBN: 978-1-003-60809-7 (ebk)

DOI: 10.1201/9781003608097

Typeset in Sabon
By Apex CoVantage, LLC

Contents

Author Biography ix
Preface x
Acknowledgments xiii

1 Introduction to Electric Vehicles **1**

Historical Evolution 1
Fundamental Principles 2
Classification of Electric Vehicles 3
1.1 *Historical Development* 4
1.2 *Basic Principles of EV Operation* 7
1.3 *Environmental Benefits of Electric Vehicles* 8

2 Types of Electric Vehicles **13**

Classification of Electric Vehicles 13
2.1 *Battery Electric Vehicles* 14
2.2 *Plug-In Hybrid Electric Vehicles* 17
2.3 *Hybrid Electric Vehicles* 19
2.4 *Fuel Cell Electric Vehicles* 21
2.5 *Comparing Propulsion Systems* 23

3 Electric Vehicle Components **26**

3.1 *Electric Motors and Drive Trains* 26
3.2 *Battery Packs and Chemistry* 30
3.3 *Power Electronics and Inverters* 34
3.4 *Regenerative Braking Systems* 37
3.5 *Vehicle Control Systems* 41

4 Battery Technology 46

4.1 Lithium-Ion Batteries 46
4.2 Nickel–Metal Hydride Batteries 49
4.3 Solid-State Batteries 51
4.4 Battery Management Systems 55
4.5 Challenges and Future Trends in Battery Technology 58

5 Charging Infrastructure 63

5.1 Home Charging Stations 63
5.2 Public Charging Stations 66
5.3 Fast-Charging Networks 69
5.4 Charging Standards and Compatibility 72

6 Range and Efficiency 77

6.1 Factors Affecting Range 77
6.2 Increasing Efficiency via Regenerative Braking 81
6.3 Energy Management Systems 84

7 Electric Vehicle Maintenance 90

7.1 Battery Care and Maintenance 90
7.2 Tire Maintenance 94
7.3 Common EV Issues and Troubleshooting 97
7.4 Maintenance Compared to Internal Combustion Engine (ICE) Vehicles 101

8 Environmental Impact 105

8.1 Carbon Emissions and Energy Sources 105
8.2 Lifecycle Assessment of Electric Vehicles 108
8.3 Comparing Environmental Footprints 111

9 Market Trends and Adoption 119

9.1 Global Market Trends in Electric Vehicles 119
9.2 Influence of Government Policies 120
9.3 Government Incentives and Policies 123

10 Future Developments 131

10.1 Advancements in Battery Technology 131
10.2 Autonomous Driving and Electric Vehicles 134

10.3 *Vehicle-to-Grid Integration* 136
10.4 *Materials and Innovations* 140

11 Case Studies 149

11.1 *Successful EV Manufacturers* 149
11.2 *Market Penetration Strategies* 153
11.3 *Real-World Adoption in Various Sectors* 156
11.4 *Case Studies From Asia* 159
11.5 *Case Studies From Europe* 163
11.6 *Case Studies From Africa* 166

12 Challenges and Solutions 171

12.1 *Expanding Charging Infrastructure: Challenges and Solutions* 171
12.2 *Battery Recycling and Sustainability: Challenges and Solutions* 175
12.3 *Affordability and Accessibility: Challenges and Solutions* 178
12.4 *Policy and Regulatory Challenges: Barriers to Scalable EV Adoption* 183

13 Safety and Standards 189

13.1 *Crash Tests and Safety Standards in Electric Vehicles* 189
13.2 *Fire Safety and Electrical Hazards in Electric Vehicles* 195
13.3 *Industry Standards and Regulations in Electric Vehicles* 201

14 Consumer Education 208

14.1 *Choosing the Right EV Model* 208
14.2 *Understanding Charging Costs* 213
14.3 *Managing Range Anxiety* 217
14.4 *Tips for Electric Vehicle Buyers* 221

15 Global Perspectives 228

15.1 *Cultural Perceptions of Electric Vehicles* 228
15.2 *Social Impacts of Electric Vehicle Adoption* 231

16 Emerging Technologies 240

16.1 Solid-State Batteries 240
16.2 Wireless Charging 245
16.3 Advanced Driver-Assistance Systems 250

17 Industry Collaboration and Partnerships 258

17.1 Collaborations Between Automakers and Tech Companies 258
17.2 Role of Startups and Innovation Hubs 262
17.3 Public–Private Partnerships 266

18 Consumer Behavior and Market Dynamics 273

18.1 Consumer Behavior and Preferences 273
18.2 Market Drivers and Barriers 279

19 Conclusion and Future Outlook 290

19.1 Key Takeaways 290
19.2 Electric Vehicles in Reducing Greenhouse Gas Emissions 294
19.3 Transforming the Automotive Industry 298

Appendices 303
A. Glossary of Terms 305
B. List of Abbreviations 306
C. Additional Resources 307
Index 309

Author Biography

Dolly Thankachan is Associate Professor and Head of the Department of Electrical and Electronics Engineering at Oriental University, Indore, India. She has published more than 56 research papers in esteemed journals, has secured six national and international patents, and contributed to four book chapters in recognized publications. Her areas of research interest include power generation, power engineering, power system analysis, control theory, MATLAB simulation, and renewable energy technologies. She is a member of professional organizations such as IEEE, IETE, and IEI.

S. Sangeetha is presently working as Assistant Professor in the Department of Electronics and Communication Engineering, Anna University, Chennai, India. Her research interests encompass microwave sensors, antennas for hand-held devices, signal integrity analysis, reconfigurable antennas, filters, microwave devices, MIMO antennas, and machine learning. She has more than 35 publications in reputable journals and contributions to four book chapters.

Shweta. B. Thomas currently serves as Assistant Professor in the School of Electronics Engineering, Vellore Institute of Technology, Vellore, India. Her research interests include digital signal processing, ground penetrating radar (GPR) signal processing, radar transmission, and detection, as well as digital and statistical signal processing for wireless physical layers, channel modeling, and wireless communication. She has published research papers in journals and conferences of national and international reputation.

Preface

The objective of an electric car book is usually clear. A glimpse of tomorrow wrapped in lithium and software, cleaner air, and novel gear are promised. However, these chapters' main tale unfolds with restless fascination. Modern electric cars are complicated systems of ideas, technology, and cultural revolutions that sometimes clash, sometimes flow elegantly. Engineers, lawmakers, battery scientists, urban designers, and drivers who are used to silence are shaping that environment. Electric mobility began over a century ago when engineers tested ideas that would mature later. Batteries, infrastructure, and patience have returned. Not out of nostalgia, this book begins at the beginning because beginnings shape future assumptions. The first part covers electric car development from prototypes to transportation ecosystem competitors. It soars during global reinvention, dips during technical stagnation, and zigs over decades of explorations. Later chapters develop the story. Electric vehicles (EVs) include battery electrics, plug-in hybrids, fuel cell, and extended-range variants. Types differ in mechanical and philosophical temperament. Transitional hybrids are preferred by some over full electrification. Understanding these contrasts is crucial as each technology attracts consumers with different performance, sustainability, and practicality needs. The second chapter sets the stage for the technical content.

Electric motors are captivating. Electric motors run quietly, unlike internal combustion engines, which yell and spit and quiver via thermodynamic acrobatics. This subtle miracle is examined in the third chapter, where electromagnetic rotation and drivetrain combinations effect efficiency and comfort. Since EVs are not just fossil fuel cars with a different power source, they need their unique engineering and appearance. Compact, powerful, and reliable electric motors offer new movement idea sets. Batteries not only power electric mobility but also cause problems. Cost, weight, charging time, range, and material ethics are set. Chapter 4 describes battery technology's miracles and messiness. Lithium-ion chemistry helped mainstream acceptance but not perfectly. It requires rare minerals, complicates recycling, and has geopolitical concerns beyond engineering. In the chapter, solid-state, lithium–sulfur, and sodium-ion batteries promise to solve difficulties but introduce new uncertainties. Infrastructure always comes up in battery discussions. Invisibly, charging stations promote the clean-energy revolution. Chapter 5 explores this patchwork network—urban hubs with many stations, rural areas without them—rapid, cheap charging, and slow, expensive, or unreliable charging. The chapter explains how consumer psychology and physical infrastructure interact: if charging is hard, individuals won't buy electric cars, and if few do, corporations won't expand the charging network. Intelligent regulations and long-term planning are needed to handle this complex feedback cycle. Longtime EV drivers know range anxiety is a spectrum of emotions. Math, memorization, topography, weather, traffic, and energy use are involved. Chapter 6 investigates range and efficiency effects of bulk, aerodynamics, surroundings, and driving style. EVs are fascinating and humble because they gently turn energy into motion without escaping thermodynamics.

So many new pieces require fresh maintenance. Chapter 7 discusses how technicians, manufacturers, and owners must adapt to a world where software updates matter as much as wrenches and diagnostics employ data streams instead of mechanical sense. Electric drivetrains have fewer moving parts, but battery heat management, inverter behavior, and regenerative braking systems demand technical knowledge. Digitally dependent, quieter, predictive maintenance sets these vehicles distinguish themselves from their internal combustion predecessors. Public conversation centers on environmental issues, which are more complex than promotional materials suggest. EVs effect mining, manufacturing, and grid composition while reducing pollution. Chapter 8 discusses nuanced intricacy. Although flawed, EVs pollute less than petroleum-based vehicles, especially when powered by renewable electricity, according to a life-cycle analysis. The chapter highlights that governments, utilities, and customers must adopt responsible policy and infrastructure for environmental gains. Technology shapes markets, and markets shape technology. Chapter 9 examines global EV adoption—who buys, why, and what impediments remain. The switch to electric mobility is driven by regional factors such economic incentives, environmental urgency, and municipal policy reforms.

In Chapter 10, innovation is anticipated. Be hopeful yet cautious about EV predictions. Solid-state batteries, bidirectional charging, and high-power quick chargers could transform EVs. Regulatory frameworks, supply-chain resilience, and customer trust affect change pace. The chapter stimulates social, economic, and technical thinking. In case studies, abstractions are grounded. Chapter 11 presents real-world installations in various countries to show large-scale EV adoption's successes, difficulties, and surprises. These stories demonstrate that people, cities, and technology collaborate on transportations. Chapter 12 emphasizes charging deserts, regulatory obstacles, and fragmented ownership. Long-distance electric road trips in certain nations have shown how unequal charging may be. Policy coordination and public–private partnerships are critiqued and encouraged in the chapter, demonstrating how some countries tackled the largest tasks. Safety is expanded in footnotes in Chapter 13. EVs bring unique fire, electrical, and accident risks that must be understood and handled. The chapter effectively argues that standardized, rigorous safety procedures will protect passengers and pedestrians and boost public confidence in a new technology. Consumer choice returns to drivers' daily lives in Chapter 14. EVs are more than just a financial or environmental choice. Personality, habits, geography, and mobility expectations are shown. This chapter helps readers understand ownership's complexities—charging patterns, energy costs, local infrastructure, and driving needs—to become informed customers and cultural change agents.

Chapter 15 investigates how technology, infrastructure, and markets affect smart mobility and sustainable city planning with electrification. Smart networks, renewable energy, and responsive cities boost EVs. The chapter shows how digital networks and environmental goals are linked to transportations. The quiet revolution of solid-state batteries begins in Chapter 16. These batteries are industry-changing because of their energy density, safety, and lifespan. Hopefully and realistically, the chapter examines solid-state system physics and engineering concerns. Chapter 17 covers artificial intelligence's (AI's) revolutionary power. Cars today are digital nodes rather than machinery. AI enables predictive maintenance, driver assistance, energy optimization, and vehicle infrastructure connectivity. This creates electric, intelligent, responsive, and autonomous transportations. In Chapter 18, consumer behavior psychology explains how emotions, habits, values, and social dynamics drive EV adoption. In unforeseen ways, environmental awareness, technological curiosity, risk aversion, and lifestyle compatibility converge. Readers are asked to examine how EVs work mechanically and in human imagination and cultures. Chapter 19 finishes with key transition insights. Electric mobility involves technological, regulatory, and behavioral changes. Today's advancements are part of a larger story of social adaptation, with lessons for manufacturers, governments, and greener transportation advocates. EVs are portrayed as engineering feats and societal experiments throughout the book chapters. It suggests advancement needs integrating several disciplines and perspectives. These chapters will show readers a calmer, more efficient, and more digitally integrated mobility environment. More innovation is needed to get broad electrification, but these pages indicate where we have been, where we are, and where we may go.

Acknowledgments

I would like to extend my deep appreciation to the many individuals who made this book happen. First, I would like to thank the authors of this volume for their outstanding chapters, which are the heart of this book. It is you and your expertise in moving electric vehicle technology forward that are truly noteworthy. I benefited immensely from the publishing house's editorial team for their guidance, patience, and professionalism throughout the complicated publication process. Their expertise clearly shaped the final manuscript. I also thank the reviewers for their careful reading and valued recommendations. Their constructive comments proved to improve the quality of the manuscript. I also owe heartfelt thanks to the many colleagues, friends, and family members for their support and understanding during the many hours dedicated to this work. Your support gave me the encouragement to finish this important endeavor. Finally, I would like to thank all the innovators and engineers around the world who are working tirelessly to ensure the transition toward sustainable mobility. This book is for them and for everyone who believes in a cleaner, greener future.

Chapter 1

Introduction to Electric Vehicles

Electric vehicles (EVs) are a futuristic breakthrough in the field of automotive technology, promising less carbon emissions, reducing dependence on fossil fuel, and probably heralding a revolution in the means of transportation. This chapter will provide a broad overview of EVs by touching on their historical evolution, essentials, and classification.

HISTORICAL EVOLUTION

The origins of EVs can be traced back to the 19th century, a time when inventors and scientists began exploring new ways to power transportation. Early experiments took inspiration from figures like Andrew Whitney, who was among the first to develop electric carriages, and Alexander-Lambert de Sivrac, who is believed to have conducted primitive experiments with electric-powered devices. These pioneers laid the groundwork, but their designs were often limited by the technology of the time. Later, inventors such as Robert Anderson built upon this early work by creating some of the first primitive electric car prototypes around 1832, using non-rechargeable batteries. Ányos Jedlik, a Hungarian inventor, also contributed to the development of electrical technology, experimenting with small electric motors that could run locomotives or small vehicles. All these efforts helped spark interest in electric power for transportation, even if early models were often impractical due to short battery life and limited range.

Despite this early activity, EVs did not gain widespread popularity for many years. It was only in the late 20th century that they began to attract serious attention from engineers and governments. This renewed interest stemmed from growing concerns over environmental pollution caused by traditional gasoline-powered cars. As urban pollution worsened and global awareness of climate change increased, the drawbacks of relying heavily on oil became more apparent. Governments and car makers started to see EVs as a clean alternative that could cut emissions and reduce dependence on

DOI: 10.1201/9781003608097-1

imported oil. Both industry leaders and environmental advocates pushed for innovations that would make electric cars more affordable, reliable, and practical. This shift was supported by advances in battery technology, such as the development of lithium-ion batteries, which helped address earlier limitations of shorter driving ranges and slow charging times. Today, EVs are seen as a vital part of efforts to reduce carbon emissions and create more sustainable transportation systems. Their history reflects a journey from early curiosity to modern solutions driven by environmental concerns and technological progress.

FUNDAMENTAL PRINCIPLES

At the core of EVs are electric motors, which serve as a direct replacement for traditional internal combustion engines (ICEs). Unlike gasoline-powered engines that burn fuel to generate power, electric motors operate using rechargeable battery packs. These batteries store electrical energy, which is then converted into mechanical power by the motor. This process allows EVs to transform electrical energy directly into kinetic energy, making their operation cleaner and more efficient.

The design of electric motors makes them highly suitable for everyday driving. They respond quickly to driver inputs, providing smooth acceleration without the noise or vibration typical of gas engines. Because electric motors rely solely on stored electric power, they eliminate or significantly reduce harmful emissions produced during operation. Compared to traditional vehicles that burn fossil fuels, EVs have a smaller carbon footprint and are considered more environmentally friendly.

The basic technology of electric motors ensures they run quietly and require less maintenance. They do not have many moving parts, which means fewer parts to wear out or break down over time. This simplicity makes EVs not only better for the environment but also easier and cheaper to keep in good condition.

Another key point is the way electric motors work during regular road use. They can be powered continuously as long as the batteries are charged, blending into normal traffic flow without causing disruption. EVs can be charged at home or at charging stations, making daily use more convenient. The ongoing shift from internal combustion engines to electric motors marks a major step toward reducing pollution and dependence on fossil fuels.

Overall, the core function of electric motors—converting electrical energy into direct motion—summarizes why EVs hold such promise for a cleaner future. Their operation is simpler and more efficient, and produces less pollution, setting them apart from traditional gasoline-powered cars. This fundamental difference shapes the future of transportation and underscores why electric motors are at the heart of the EV revolution.

CLASSIFICATION OF ELECTRIC VEHICLES

EVs encompass a diverse range of designs and configurations, each catering to specific use cases and consumer preferences. Table 1.1 provides a taxonomy of EVs based on their propulsion systems and energy sources.

As per Table 1.1, battery electric vehicles (BEVs) depend on the electricity stored in battery packs high in capacity for power generation. This eliminates tailpipe emissions entirely, making BEVs a popular choice among environmentally conscious consumers. However, their limited range and longer charging times compared to those of conventional vehicles pose practical challenges toward widespread adoption.

Plug-in hybrid electric vehicles (PHEVs) combine electric and gasoline-powered or diesel-powered engines, offering consumers the mix of energy sources. These vehicles may operate without any electricity for less than short distances and will run on gasoline or diesel when the battery gets run down. PHEVs fill the gap between conventional ICE vehicles and fully electric alternatives, providing a transitional solution for those hesitant to embrace pure electrification.

The parallel hybrid system employed by hybrid electric vehicles (HEVs) fuses the internal combustion engine with an electric motor. Unlike PHEVs, a rechargeable external power source is not an option, and regenerative braking is used to refill the battery. Although more efficient in fuel use and with less emission, HEVs are still fuel-based and thus cannot completely tame down its carbon footprint.

To summarize, this section discusses EVs—their historical importance, how they work, and the types of vehicles. It explains how the different EV types feature characteristics and benefits, which bring out the advantages of such vehicles. With such explanations on the platforms provided, readers should be aware of the landscapes in the world of sustainable transport.

Table 1.1 Taxonomy of EVs Based on Their Propulsion Systems and Energy Sources.

Type of EV	*Description*
Battery electric vehicle	Exclusively powered by rechargeable battery packs, BEVs eliminate the need for internal combustion engines
Plug-in hybrid electric vehicle	Combine electric propulsion with an internal combustion engine, offering flexibility in energy sources
Hybrid electric vehicle	Incorporate both an internal combustion engine and an electric motor, utilizing regenerative braking for energy recovery

1.1 HISTORICAL DEVELOPMENT

The development of EVs was a long-term and complicated process that reached from the earliest times in history till the beginning of the 21st century, consisting of several milestones, innovations, and technological advancements. The section below chronicles the development of EVs from their conception to the present day, shedding light on the key breakthroughs and noted contributions.

19th-Century Pioneers:

1. Robert Anderson (1832): English inventor Robert Anderson is considered one of the earliest architects of electric automobiles. Although Anderson built a simple carriage driven for a few miles at the end of the 19th century, his invention was the progenitor of future advancements in EV technology.
2. Ányos Jedlik (1828–1834): Hungarian physicist Ányos Jedlik initiated work on an electric motor-driven vehicle in 1828, preceding Robert Anderson. The EV had been built as early as 1831, and the founder of the company attached his name to the utilization of rechargeable batteries for propulsion. The contributions of Jedlik in electromagnetism and electric propulsion laid the groundwork for the progress of electric transport.

Early 20th-Century Innovations:

3. Thomas Parker (1884): Inventor Thomas Parker is a British inventor who constructed one of the first motor vehicles in terms of practicality. Parker's EVs were widely popularized in cities because they lacked an exhaust and were super quiet.
4. Milburn Electric Company (1914–1923): The Milburn Electric Company manufactured electric automobiles of the early 20th century, primarily in cities with a great population and for the needs of affluent customers and urban commuters. They introduced the Milburn Electric Model 27 in 1915, which helped make a reputation for themselves.

Mid-20th-Century Decline:

5. Rise of Internal Combustion Engines: The widespread adoption of ICE vehicles due to abundant petroleum reserves and advancements in automotive manufacturing started drifting away from the increased interest in EVs in the 1960s and 1970s. The convenience of long-distance travel and the availability of cheap gasoline also contributed to the decline of EVs during this period.
6. Revival of Interest (1960s–1970s): Environmental concerns and oil crises in the 1960s and 1970s brought back interest in EVs as a sensible alternative to traditional cars. Projects such as the General Motors

Electrovair and the AMC Amitron, although not very successful in terms of commercial application, marked great developments in electric propulsion systems.

21st-Century Renaissance:

7. Battery Technology Development: The 21st century clearly brought forward major steps in the development of batteries, especially the refinement of lithium-ion batteries. These high-energy-density batteries changed the EV industry on two sides: longer driving range, slower charging time, and much better performance.
8. Mass-Market EVs: Companies such as Tesla, Nissan, and Chevrolet brought the mass market on board EVs with releases like the Tesla Roadster, Nissan Leaf, and Chevrolet Volt. Innovative design, cutting-edge technology, and improved affordability allowed for the growing recognition of electric transportation among most consumers.

As per Table 1.2,

- Robert Anderson (1832): Although Anderson's creation is one of the first handmade, and indeed very few, electric cars, its electric carriages were indeed an early step toward electric automobiles. It was, however, hindered by the battery technology at the time of its creation, but it underscored the practicality of electric propulsion for the movement of people and goods.

Table 1.2 Milestones in the Historical Development of EVs.

Milestone	*Description*
Robert Anderson (1832)	Creation of one of the earliest electric carriages powered by non-rechargeable primary cells
Ányos Jedlik (1828–1834)	Development of a prototype electric motor-driven vehicle, laying the foundation for future EVs
Thomas Parker (1884)	Construction of practical electric cars using rechargeable batteries for propulsion
The Milburn Electric (1914–1923)	Production of electric automobiles with significant range and popularity among urban commuters
Rise of ICEs	Dominance of ICE vehicles due to cheap gasoline and convenience for long-distance travel
Resurgence of interest (1960s–1970s)	Renewed interest in EVs amid environmental concerns and oil crises
Advancements in battery technology	Breakthroughs in lithium-ion batteries enhancing EV performance and driving range
Introduction of mass-market EVs	Release of affordable and technologically advanced EV models driving mainstream adoptions

- Ányos Jedlik (1828–1834): Ányos Jedlik is a pioneering figure who did pioneering work on a prototype electric motor-driven vehicle, thereby laying the groundwork for later developments in electric transportation. Furthermore, such experiments opened doors to future advancements in electric propulsion.
- Thomas Parker (1884): Thomas Parker's development of practical EVs, using rechargeable batteries, popularized EVs in the closing decades of the 19th century. Parker's advances in battery technology and electric motor design worked toward the viability of electric transportation as an alternative to ICE vehicles.
- Milburn Electric (1914–1923): The Milburn Electric Company's production of electric automobiles, marketed with significant range and popularity among the urban commuters, brought home the fact that the EV was possible. Model 27, in particular, manifested advances in battery technology and the design of vehicles.
- Rise of ICEs: The more widespread dominance of ICE vehicles in the mid-20th century overshadowed the popularity of electric cars due to cheap gasoline and the convenience of long-distance travel. In spite of the early successes, EVs were faced with stiff competition from their gasoline-powered counterparts.
- Revival of Interest (1960s–1970s): The environmental concerns that arose with the advent of the 1960s and 1970s, coupled with oil crises, renewed interest in EVs as alternatives to conventional automobiles. This set of experimental projects meant there was the potential of electric propulsion systems to counter the negative environmental effect and dependence on fossil fuels.
- Battery Technology Breakthroughs: In the 21st century, there has been significant progress in lithium-ion battery technology, which changed the EV market entirely. The advancements were in terms of the full range of the car, faster charging that has come from the technology, and improvements in overall performance that would lead to the mass market adopting EVs worldwide.
- Entry of Mainstream EVs: The introduction of mass-market EV models by companies such as Tesla, Nissan, and Chevrolet marked a major milestone in mainstream acceptance of electric transportation. Mass-produced electric cars were affordable and technologically advanced, with a better performance, range, and convenience level compared to conventional ICE vehicles.

Therefore, this historic development of EVs comprised significant milestones, innovations, and technological advancements in terms of the progressive development of electric transportation in all aspects. Through these, the history of the development of EVs will remain a valuable guide for both developers and users on the evolution of EVs into sustainable transportation solutions.

1.2 BASIC PRINCIPLES OF EV OPERATION

EVs operate on fundamentally different principles from traditional ICE vehicles. This section elucidates the core principles governing the operation of EVs, from energy conversion mechanisms to propulsion systems.

Energy Conversion Mechanisms

At the heart of EV operation lies the conversion of electrical energy into kinetic energy. This process occurs through the interaction among various components, including the battery pack, electric motor, power electronics, and regenerative braking system.

Table 1.3 delineates the key components involved in the energy conversion process within an EV.

As per Table 1.3,

1. Battery Pack: A battery pack is an energy storage device that converts and stores electric energy in the chemical form in an EV. As the vehicle operates, the battery supplies power to the electric motor, facilitating propulsion.
2. Electric Motor: Electric motors converts electrical energy to mechanical energy, which propels the wheels. The transfer of combustion processes used in conventional gasoline automobiles is different, with electric motors providing quick acceleration and instant torque to enhance driving manners.
3. Power Electronics: Power electronics control the flow of energy between the battery and the electric motor. Such components regulate aspects of electrical energy, such as voltage, current, and frequency, to achieve good performance and efficiency from the propulsion system.

Table 1.3 Components of EV Energy Conversion.

Component	*Function*
Battery pack	Stores electrical energy in chemical form, providing power to the electric motor for propulsion
Electric motor	Converts electrical energy from the battery into mechanical energy, propelling the vehicle forward
Power electronics	Controls the flow of electricity from the battery to the electric motor, regulating speed and torque output
Regenerative braking system	Recaptures kinetic energy during braking, converting it back into electrical energy to recharge the battery

4. Regenerative Braking System: Typically, the dissipation of kinetic energy by conventional vehicles during braking is in heat form. However, EVs use the regenerative braking system to regenerate the energy dissipated while braking into electrical energy. This energy is then fed back to the battery and enhances the efficiency and driving range of the EV.

Readers get an insight into how it functions and how these work together to make an EV perform as they move on to the subsequent discussion of modernizations in EV technology.

1.3 ENVIRONMENTAL BENEFITS OF ELECTRIC VEHICLES

EVs offer a promising solution to mitigate the environmental impacts associated with conventional ICE vehicles. This section delves into the environmental benefits of EVs, elucidating their role in reducing greenhouse gas (GHG) emissions, improving air quality, and conserving natural resources.

Reduction in Greenhouse Gas Emissions

One of the primary environmental advantages of EVs lies in their significantly lower emissions of GHGs compared to ICE vehicles. This reduction stems from the absence of tailpipe emissions in BEVs and the overall higher efficiency of electric drivetrains.

Table 1.4 quantifies the reduction in GHG emissions achieved by EVs compared to their ICE counterparts.

As per Table 1.4,

1. BEVs: BEVs emit zero tailpipe emissions during the use process and therefore have 100% reduction of the GHG emissions compared to the ICE vehicles. Such a reduction in these emissions greatly improves air quality and assists in mitigating the impacts of climate change.

Table 1.4 Reduction in GHG Emissions with EV Adoption.

Vehicle Type	*GHG Emissions Reduction (%)*
BEV	100% (Zero tailpipe emissions)
PHEV	50–75% (Dependent on driving mode and battery usage)
HEV	25–50% (Dependent on driving conditions and regenerative braking effectiveness)

2. PHEVs: PHEVs have lower GHG emissions compared to conventional ICE vehicles, but the degree of reduction differs depending on the driving mode and battery usage. When operating in electric-only mode, PHEVs realize emission reductions in line with those of BEVs. However, when using the ICE, emissions are in line with the emissions of the ICE.
3. HEVs: Although HEVs incorporate electric propulsion and regenerative braking systems to improve efficiency, they still rely on ICEs for primary propulsion. Due to this, the reduction in GHG emissions is moderate compared to that of BEVs and PHEVs, with variability based on the driving conditions and the effectiveness of regenerative braking.

Improvement of Local Air Quality: Improvements in air quality are also attained as a result of the adoption of EVs, as it eliminates tailpipe pollutants such as nitrogen oxides (NOx), particulate matter (PM), and volatile organic compounds (VOCs). Table 1.5 illustrates the reduction in air pollutants achieved through EV deployment.

As per Table 1.5,

1. Nitrogen Oxides: NOx are harmful pollutants produced mainly from burning fossil fuels in engines that use internal combustion. Traditional gasoline or diesel cars release significant amounts of NOx into the air through their tailpipes. These gases play a major role in forming smog, which can cause breathing problems and worsen lung diseases. NOx also make ground-level ozone, a key part of smog, more intense, leading to hazy skies and unhealthy air conditions. People living near busy roads or industrial areas are often exposed to high levels of NOx, increasing their health risks.

   EVs do not have exhaust pipes like gas-powered cars. Because they run on electricity, they produce no tailpipe emissions of NOx. This means EVs do not directly contribute to NOx pollution. As more EVs are used, especially in large cities, the amount of NOx released into the atmosphere drops. This can help improve air quality noticeably in urban areas, leading to clearer skies and healthier lives. Experts say

Table 1.5 Reduction in Air Pollutants With EV Adoption.

Pollutant	*Reduction (%)*
NOx	100% (Zero tailpipe emissions)
PM	100% (Zero tailpipe emissions)
VOCs	100% (Zero tailpipe emissions)

that replacing traditional cars with EVs could cut NOx emissions by a large margin. For example, in cities with many electric cars, the level of NOx pollution has decreased significantly.

Stopping NOx from entering the air is crucial for health and environmental reasons. Reducing NOx also helps slow down how quickly smog forms. During sunny days, NOx reacts with sunlight to create ground-level ozone, which can cause coughing and throat irritation and worsen asthma. If EVs become more common, the overall production of this dangerous pollutant decreases. This shift also benefits asthma patients and children, who are more vulnerable to respiratory problems caused by polluted air. Experts agree that EVs are a key part of cleaning up urban air. They stress that reducing NOx emissions directly benefits public health and the environment.

In sum, EVs play an important role in fighting air pollution. They produce no NOx during operation. Their widespread use can cut urban smog and improve air quality. This reduction helps protect everyone's health while making cities cleaner and safer places to live.

2. Particulate Matter: PM is made up of tiny particles that float in the air. These particles come from many sources, especially from vehicles that burn fuel. When a car runs, its tailpipe releases these small particles into the air. These particles are so tiny that they can easily be inhaled into the lungs. Scientists link exposure to PM with health problems like asthma, lung infections, and heart disease. High levels of PM are common in cities with many cars and trucks. These particles harm air quality and make it harder for people to breathe.

   EVs change this situation. Unlike gasoline-powered cars, EVs do not have tailpipes. Since they run on electricity, they produce no direct emissions related to PM. This leads to a 100% reduction in PM emissions from these vehicles. When more people switch to EVs, cities experience cleaner air. This is especially important for communities near busy roads or factories. Reduced PM levels can help improve long-term health for residents.

   Cleaner air benefits everyone, but it especially helps children, the elderly, and people with existing health problems. They are more vulnerable to the effects of tiny particles in the air. Experts see EVs as a way to cut air pollution significantly. For example, cities that promote electric cars tend to see lower pollution levels and fewer health issues. In the future, the wider use of EVs could lead to much healthier communities and a better environment.

3. VOCs: EVs have a significant environmental advantage because they do not release VOCs during their operation. VOCs are chemicals that easily become vapors or gases. They come from sources like gasoline engines and fuel combustion. When traditional gasoline cars run, they emit VOCs into the air, which can harm air quality. These emissions

contribute to the formation of ground-level ozone, a major part of smog that affects health and the environment.

Unlike gasoline-powered cars, EVs run on electricity. This means they do not burn fuel in the engine, so they do not produce VOCs during normal driving. As a result, EVs help keep the air cleaner. They also play a role in reducing smog, which is made worse by these harmful chemicals. Clearer air reduces health problems such as asthma and respiratory issues for millions of people.

The impact of EVs on air quality is especially important in cities. Urban areas often face high levels of pollution from traffic. When more EVs are used, emissions of VOCs drop, helping cities meet clean air standards. Experts say that replacing traditional vehicles with EVs can cut VOC emissions significantly. Many environmental groups encourage this switch because of its positive effect on air health.

In addition, reducing VOC emissions from EV use lowers the formation of ground-level ozone. Ozone, while beneficial high in the sky, becomes dangerous at the ground level. It can cause breathing problems and other health issues. Cutting down VOCs means less ozone will form near the ground, making the air safer to breathe. This benefit is especially critical during warmer months when sunlight speeds up pollution reactions.

Overall, EVs offer a cleaner alternative that does not contribute to the release of volatile organic compounds. This change supports better air quality and reduces the harmful smog created when VOCs and other pollutants react with sunlight. As the world moves toward cleaner transportation, EVs stand out for their ability to operate without adding these harmful chemicals to the atmosphere. This is a key reason why many cities and countries are investing heavily in EV adoption.

Quantifying the environmental benefits of EVs shows how they can significantly cut down GHG emissions. Transportation accounts for a large share of emissions that warm the planet. Switching from traditional gasoline-powered cars to EVs can reduce these emissions by a substantial margin. For example, studies show that EVs produce up to 60% fewer emissions over their lifetime than conventional cars. This reduction helps slow down climate change, which causes extreme weather events and rising sea levels.

EVs also improve air quality in cities. Gas-powered cars release pollutants like NOx and PM that can harm people's lungs. These pollutants cause health problems, such as asthma and lung disease. In contrast, EVs produce no tailpipe emissions. When powered by cleaner energy sources like wind or solar, they produce even fewer pollutants. Cities that promote EV use see a drop in smog and better air quality for residents. For example, some urban areas that shifted to electric taxis experienced a notable drop in local air pollution.

This section highlights the important role that EVs can play in moving toward greener transport systems. Transitioning to EVs is key to reducing our carbon footprint. It also encourages governments and industries to invest in cleaner energy and charging infrastructure. This shift can help many countries meet the climate goals they have set. It's clear that widespread EV adoption has the potential to make a real difference.

The urgency to speed up EV adoption cannot be overstated. As climate change accelerates, the need for cleaner transportation options becomes more urgent. Promptly increasing EV use reduces reliance on fossil fuels and lowers emissions quickly. It also supports healthier living conditions in urban areas. Experts agree that faster adoption of EVs will have a positive impact on overall public health. It is a vital step toward protecting future generations from severe climate and health risks.

Making the switch to EVs is no longer just a choice: it's a necessity. Governments, businesses, and individuals all have a role to play in this effort. By doing so, we can reduce the damage caused by traditional transportation methods. The faster we adopt EVs, the better we can combat climate change and improve air quality for everyone. Transitioning to EVs is essential for a more sustainable and healthier future.

Chapter 2

Types of Electric Vehicles

Electric vehicles (EVs) take forms ranging from simple motor-only carts to complex systems using up-to-date technologies. This chapter gives an overview of all the major forms of EVs, listing their distinguished features and propulsion systems, with a view to the applications in which they are mostly used.

CLASSIFICATION OF ELECTRIC VEHICLES

EVs are classified into different categories based on the type of propulsion system, mode of energy, and features of operation. Table 2.1 presents the taxonomy of EVs, which helps classify all the major forms of EVs.

As per Table 2.1,

1. Battery Electric Vehicles (BEVs): BEVs represent the purest form of EVs, relying exclusively on electric propulsion powered by rechargeable battery packs. These vehicles emit zero tailpipe emissions during operation, making them an environmentally friendly option for urban commuting and short- to medium-range travel. BEVs are well-suited for drivers with access to charging infrastructure and those seeking to minimize their carbon footprint.
2. Plug-In Hybrid Electric Vehicles (PHEVs): PHEVs combine electric propulsion with an internal combustion engine (ICE), offering drivers the flexibility to operate in electric-only mode or utilize the combustion engine for extended range. PHEVs feature rechargeable battery packs that can be charged from external power sources, enabling emission-free driving for short distances while providing the option for longer trips without range anxiety. This versatility makes PHEVs an attractive choice for consumers transitioning to electric mobility.
3. Hybrid Electric Vehicles (HEVs): HEVs integrate both an ICE and an electric motor, utilizing regenerative braking to recharge the battery during deceleration. Unlike PHEVs, HEVs cannot be recharged

DOI: 10.1201/9781003608097-2

Table 2.1 Taxonomy of EVs Based on the Major Forms.

Type of EV	*Description*
Battery electric vehicle	Propelled solely by electric motors powered by rechargeable battery packs, with zero tailpipe emissions
Plug-in hybrid electric vehicle	Combines electric propulsion with an internal combustion engine, offering flexibility in energy sources
Hybrid electric vehicle	Incorporates both an internal combustion engine and an electric motor, utilizing regenerative braking for energy recovery
Fuel cell electric vehicle	Utilizes hydrogen fuel cells to generate electricity, offering long driving ranges and rapid refueling times

externally and rely primarily on gasoline or diesel fuel for propulsion. While offering improved fuel efficiency and reduced emissions compared to conventional vehicles, HEVs represent a transitional technology toward full electrification.

4. Fuel Cell Electric Vehicles (FCEVs): FCEVs utilize hydrogen fuel cells to generate electricity through an electrochemical reaction, powering electric motors for propulsion. These vehicles offer long driving ranges and rapid refueling times, making them suitable for long-distance travel and heavy-duty applications. FCEVs emit only water vapor as exhaust, making them a zero-emission alternative to traditional ICE vehicles. However, limited hydrogen infrastructure and production challenges currently restrict their widespread adoption.

By delineating the characteristics and applications of various types of EVs, this section provides readers with a comprehensive understanding of the evolving landscape of sustainable transportation. Each vehicle type offers unique advantages and considerations, catering to diverse consumer preferences and driving needs in the transition toward a greener automotive future.

2.1 BATTERY ELECTRIC VEHICLES

BEVs are more technologically advanced cars that run on charge from a large battery pack in full electric operation. This part explores BEVs in detail—their basic principles of operation, benefits, disadvantages, and critical factors for consumers.

Operational Principles

The basic operation of BEV is by the use of electric motors powered by rechargeable battery packs. When the vehicle is in motion, electrical energy stored in the battery is converted into mechanical energy by an electric motor, thereby propelling the vehicle forward. Compared to ICEs that use the burning of fossil fuels, BEVs emit zero tailpipe emissions during operation and therefore do a great job in improving air quality and reducing greenhouse gases.

Advantages of BEVs

BEVs have several advantages compared to conventional ICE vehicles and other types of EVs. They include the following:

1. Zero Emissions: BEVs do not emit any tailpipe emissions, thereby reducing air pollution and hence the environmental impact of driving.
2. Quiet Operation: Electric motors work silently, compared to the noisy ICEs, so the noise levels are reduced, and a quieter ride is maintained.
3. Lower Operating Costs: BEVs have fewer moving parts than ICE vehicles, resulting in lower maintenance and repair costs over their lifetime.
4. Energy Efficiency: EVs are inherently more energy efficient than combustion engines, resulting in a larger amount of energy per mile.

Challenges and Considerations

BEVs may face some challenges and considerations that might be a barrier to their wider use:

1. Limited Range: The battery pack is the size of the one used in vehicles, and because of it, the range of BEVs is normal—people can worry about the fact that they won't reach a distant place because their car has run out of charge.
2. Charging Infrastructure: One of the main problems for the propagation of BEVs is the lack of charging infrastructure, as is especially true for rural areas.
3. Charging Time: Although there are quick-charging technologies, a BEV still requires more time to recharge compared to refueling conventional vehicles, causing inconvenience to some drivers.
4. Battery Degradation: Over time, the battery performance and capacity of BEVs degrade, thus reducing the driving range and increasing charging frequency.

Table 2.2 shows the following:

- Emissions: BEVs fail to emit any tailpipe emissions, making them the most environmentally friendly of all the listed vehicle types. PHEVs and HEVs emit fewer emissions compared to conventional ICE vehicles, although their environmental impact varies with the driving mode and fuel usage.
- Range: BEVs have a limited range determined by the size of the battery pack, while PHEVs can supplement electric range by the use of an ICE. In contrast, HEVs and ICE vehicles are solely dependent on fuel tanks, with range depending on the fuel capacity and fuel efficiency.
- Charging: BEVs need access to charging infrastructure for recharging, whereas PHEVs can operate in the hybrid mode without external charging. HEVs and ICE vehicles do not need external charging facilities, relying solely on refueling at gas stations.
- Maintenance: BEVs have fewer moving parts compared to ICE vehicles; thus, their maintenance is the lowest over the vehicle's lifespan. PHEVs and HEVs require moderate maintenance; at the same time, ICE vehicles are prone to routine maintenance due to the ICE.

This section would enable the reader with a comprehensive understanding of this transformative vehicle technology. Although this technology is faced with challenges such as the limitation of their range and inadequate charging infrastructure, it represents a sustainable and environmentally responsible solution for personal and commercial transportation needs.

Table 2.2 Comparison of BEVs with Other Vehicle Types.

Aspect	*BEV*	*PHEV*	*HEV*	*ICE Vehicle*
Emissions	Zero tailpipe emissions	Emissions vary based on the driving mode	Reduced emissions	Emissions from combustion of fossil fuels
Range	Limited by battery capacity	Electric range supplemented by ICEs	Dependent on fuel tank capacity	Dependent on fuel tank capacity
Charging	Requires access to charging infrastructure	Can operate in hybrid mode without charging	No external charging required	Refueling at gas stations
Maintenance	Fewer moving parts, lower maintenance costs	Moderate maintenance requirements	Moderate maintenance requirements	Routine maintenance required
Fuel/power source	Electricity	Electricity and gasoline/ diesel	Gasoline/diesel	Gasoline/diesel

2.2 PLUG-IN HYBRID ELECTRIC VEHICLES

PHEVs are a class of vehicles that represent a flexible alternative to both traditional ICE vehicles and BEVs. This section will elaborate on the operational principles of PHEVs, their advantages, their challenges, and major considerations for consumers.

Operational Principles

A combination of electric propulsion and an ICE can be observed in PHEVs. Rechargeable battery packs are electrically produced to be recharged from an external power source, which can work for a certain range. On exhaustion of the battery or in the event that any other reason necessitates an additional source of power, the vehicle makes a comfortable transition to utilize the ICE, resulting in a longer driving range and thereby eliminating the fear of running out of battery during long trips.

Advantages of PHEVs

PHEVs have many advantages over the traditional ICE vehicles and BEVs; some of these include the following:

1. Flexibility: With PHEVs, the driver can run on electric-only mode for short-distance use and the ICE for longer trips, free from range anxiety.
2. Reduced Emissions: PHEVs run on electric-only mode emit no tailpipe emissions, resulting in improvement in air quality and lowering of greenhouse gas emissions. Even when operating in the hybrid mode, PHEVs emit less than conventional ICE vehicles.
3. Longer Range: The combination of electric propulsion and an ICE allows PHEVs to offer longer driving ranges compared to BEVs, hence making them a choice for a broader set of driving and travel needs.
4. Flexibility on Charging: Charging of PHEVs can be done via external power sources such as home-based charging stations or public charging infrastructures, providing drivers a convenient way of charging anywhere.

Challenges and Considerations

While PHEVs have their advantages, they also present challenges and considerations that may impact their adoption and usage:

1. Electric-Only Range: The range of PHEVs in the electric mode is usually limited by the capacity of the battery pack, requiring a switch to the hybrid mode or ICE use for longer trips.

2. Complexity: PHEVs integrate dual propulsion systems (electric motor and ICE); therefore, such vehicles may come with more complexity compared to conventional ICE vehicles and maintenance necessities.
3. Charging Infrastructure: PHEVs, although they offer the affordability of electric propulsion, still have to contend with the availability of charging infrastructures to recharge their battery packs.
4. Cost: Compared to conventional ICE vehicles, PHEVs are typically more expensive because they are based on the complexity of the dual propulsion systems and rechargeable battery pack. However, with the savings in fuel use and the various government incentives, the initial cost would have been offset by now.

As per Table 2.3,

1. Emissions: PHEVs produce zero tailpipe emissions while traveling in all-electric mode and reduced emissions in the hybrid mode; hence, they are more environment-friendly than the traditional ICE vehicles. BEVs produce zero tailpipe emissions, whereas HEVs emit fewer emissions than ICE vehicles.

Table 2.3 Comparison of PHEVs With Other Vehicle Types.

Aspect	*PHEV*	*BEV*	*HEV*	*ICE Vehicle*
Emissions	Zero tailpipe emissions in electric mode, reduced emissions in hybrid mode	Zero tailpipe emissions	Reduced emissions	Emissions from combustion of fossil fuels
Range	Electric range supplemented by ICE, longer total range compared to BEVs	Limited by battery capacity	Dependent on fuel tank capacity	Dependent on fuel tank capacity
Charging	Requires access to charging infrastructure for electric mode	Requires access to charging infrastructure	No external charging required	Refueling at gas stations
Flexibility	Operates in electric-only mode or hybrid mode, providing flexibility for different driving needs	Limited by battery range	Operates using both electric and ICE propulsion	Solely relies on ICE

2. Range: PHEVs have a combined electric-only and gasoline range, offering a longer total driving range compared to BEVs with a battery capacity. BEVs are limited by battery capacity, while HEVs and ICE vehicles rely on gasoline tanks only for range.
3. Charging: PHEVs require the ability to recharge the battery while the car is in motion, so their charging infrastructure differs from what BEVs require. Charging is needed for HEVs and ICE vehicles through gas stations.
4. Flexibility: PHEVs come with an option of powering up in either the electric or hybrid mode, hence ensuring that the driver gets flexibility concerning his range or driving demands. In return, BEVs are limited in terms of battery range, while HEVs and ICE vehicles rely solely on ICEs for transportation.

This section delineates the operational principles, advantages, challenges, and considerations of PHEVs, providing readers with a comprehensive understanding of this hybrid vehicle technology. PHEVs represent an option for consumers who want to move toward a greener future but don't want to leave behind the hybrid propulsion's flexibility and convenience.

2.3 HYBRID ELECTRIC VEHICLES

HEVs exemplify an intermediate step toward electrification, integrating the benefits of electric propulsion and the inherent capability of an ICE. This section explores the full operation of HEVs, which include its ways of operation, advantages, weaknesses, and essential considerations to the consumer. Operational Principles

Parallel hybrid systems allow HEVs to integrate an electric motor and an ICE for propulsion. During acceleration and cruising, the electric motor assists the ICE, cutting fuel use and emissions. In addition, HEVs use a regenerative braking system, capturing kinetic energy during deceleration and converting it into electrical energy to recharge the battery pack.

Advantages of HEVs

HEVs have several benefits over conventional ICE vehicles, including the following:

1. Improved Fuel Efficiency: Incorporating electric propulsion and regenerative braking systems improves the overall fuel efficiency of HEVs compared to traditional ICE vehicles, thus decreasing fuel consumption and emissions.
2. Reduced Emissions: While not as environmentally friendly as BEVs or PHEVs, HEVs result in fewer emissions compared to traditional ICE

vehicles, thus contributing to improved air quality and reduced environmental impact.

3. Enhanced Driving Dynamics: The electric motor in HEVs provides instant torque and smoother acceleration compared to traditional ICE vehicles, hence improving driving dynamics and overall performance.
4. No Range Anxiety: Unlike BEVs that can have a limit on the battery range, HEVs are based on an ICE and are not entirely dependent on electric propulsion to avoid range anxiety for longer trips.

Challenges and Considerations

Despite their advantages, HEVs also face challenges and considerations that may limit their take-up and usage:

1. Limited Electric-Only Operation: HEVs cannot function like BEVs or PHEVs, relying mostly on the electric motor for propulsion. While regenerative braking systems facilitate battery recharging, HEVs have zero operation in electric-only mode.
2. Complexity and Maintenance: HEVs are double-propelled systems, employing an electric motor and an ICE, where the complexity is increased over conventional ICE vehicles. However, since HEVs do not have a rechargeable battery pack, the requirements for maintenance with respect to PHEVs are almost nil.
3. Environmental Impact: Despite reduced emissions compared to conventional ICE vehicles, HEVs still use fossil fuels as a source of energy for propulsion, resulting in the emission of greenhouse gases and air pollution. Although the environmental impact of HEVs is less than that of traditional ICE vehicles, it is non-zero in the sets.
4. Cost: HEVs are usually more costly than conventional ICE vehicles due to dual propulsion systems and hybrid technology. However, the potential fuel savings over the life of a car can offset this difference for some buyers.

Table 2.4 details the following:

1. Emissions: While utilizing less fuel, conventional ICE vehicles emit fewer emissions than HEVs, contributing positively to air quality and reduced environmental impact. BEVs and PHEVs have zero tailpipe emissions in the electric mode, while PHEVs have lower emissions in the hybrid mode.
2. Range: Refueling is essential for HEVs' driving ranges since the ranges are set by fuel tank capacity and fuel efficiency. BEVs have an electric range associated with the battery capacity, while PHEVs combine the two, returning flexibility for hybrid driving.
3. Charging: HEVs do not require external charging infrastructures but rely on gas station refueling only. BEVs and PHEVs, on the other hand,

Table 2.4 Comparison of HEVs With Other Vehicle Types.

Aspect	*HEV*	*BEV*	*PHEV*	*ICE Vehicle*
Emissions	Reduced emissions compared to ICE vehicles	Zero tailpipe emissions	Zero tailpipe emissions in electric mode, reduced emissions in hybrid mode	Emissions from combustion of fossil fuels
Range	Dependent on fuel tank capacity	Limited by battery capacity	Electric range supplemented by ICE, longer total range compared to BEVs	Dependent on fuel tank capacity
Charging	No external charging required	Requires access to charging infrastructure	Requires access to charging infrastructure for the electric mode	Refueling at gas stations
Fuel efficiency	Improved compared to ICE vehicles	No fuel consumption	Variable fuel consumption depending on the driving mode	Dependent on fuel efficiency

require access to charging infrastructure for battery pack recharging in the electric mode.

4. Fuel Efficiency: Conventional ICE vehicles show better fuel efficiency compared to other vehicles, meaning better fuel consumption in general, and in turn better greenhouse gas emissions. BEVs do not require fuel, so there is zero tailpipe emission; at the same time, the fuel consumption of PHEVs is increased depending on the driving mode.

Thus, the various operational characteristics, advantages, challenges, and considerations regarding HEVs have been explained in detail throughout the section so as to give the reader a greater understanding of the technology of hybrid vehicle operation. In essence, HEVs represent a balance among fuel efficiency, emissions reduction, and range for consumers who wish to relocate their driving toward more sustainable transportation goals.

2.4 FUEL CELL ELECTRIC VEHICLES

FCEVs represent an innovative approach to zero-emission transportation, utilizing hydrogen fuel cells to generate electricity for propulsion. This section provides an in-depth exploration of FCEVs, encompassing their operational principles, advantages, challenges, and key considerations for consumers.

Operational Principles

At the heart of FCEV operation lies the hydrogen fuel cell, which electrochemically combines hydrogen and oxygen from the air to produce electricity, water vapor, and heat. The chemical reaction that occurs within the fuel cell is represented by the following equation:

$$2H_2 + O_2 \rightarrow 2H_2O + Energy$$

In this equation:

- H_2 represents hydrogen molecules.
- O_2 represents oxygen molecules.
- $H_2\,O$ represents water molecules.
- "*Energy*" represents the electrical energy generated by the fuel cells.

The electricity produced by the fuel cell is used to power electric motors, which propel the vehicle. The only byproduct of this process is water vapor, making FCEVs a truly zero-emission transportation solution.

Explanation of the Equation

1. Hydrogen : Hydrogen molecules (H_2) are supplied to the anode side of the fuel cell.
2. Oxygen : Oxygen molecules (O_2) from the air are supplied to the cathode side of the fuel cell.
3. Water: The chemical reaction between hydrogen and oxygen within the fuel cell produces water molecules (H_2O) as a byproduct.
4. Energy Generation: The chemical reaction releases energy, which is converted into electrical energy to power the vehicle's electric motor.

Advantages of FCEVs

FCEVs offer several advantages over traditional ICE vehicles and BEVs, including the following:

1. Zero Emissions: FCEVs produce only water vapor as exhaust, making them a truly zero-emission transportation solution.
2. Long Driving Range: FCEVs offer longer driving ranges compared to BEVs, as hydrogen can be stored at higher energy densities than batteries.
3. Rapid Refueling: Refueling FCEVs with hydrogen takes a similar amount of time as refueling ICE vehicles, providing a familiar and convenient experience for drivers.
4. Scalable Technology: Hydrogen fuel cell technology is scalable and can be adapted for various vehicle types, including passenger cars, trucks, buses, and even trains.

Challenges and Considerations

Despite their advantages, FCEVs also face challenges and considerations that may impact their widespread adoption:

1. Hydrogen Infrastructure: The availability of hydrogen refueling infrastructure is limited, particularly outside of certain regions and metropolitan areas.
2. Cost: FCEVs tend to be more expensive than traditional ICE vehicles and BEVs due to the high cost of fuel cell technology and hydrogen production.
3. Hydrogen Production: The majority of hydrogen is currently produced from fossil fuels, leading to greenhouse gas emissions unless renewable energy sources are utilized in the production process.
4. Safety Concerns: Hydrogen is highly flammable and requires careful handling and storage, raising safety concerns for both consumers and infrastructure operators.

By elucidating the operational principles, advantages, challenges, and considerations of FCEVs, this section provides readers with a comprehensive understanding of this innovative vehicle technology. FCEVs offer a promising solution for achieving zero-emission transportation and reducing dependence on fossil fuels, but their widespread adoption hinges on overcoming the challenges related to infrastructure, cost, and safety.

2.5 COMPARING PROPULSION SYSTEMS

Choosing the right propulsion system is crucial for consumers considering EVs as their mode of transportation. This section provides a comprehensive comparison of various propulsion systems, including BEVs, PHEVs, HEVs, and ICE vehicles. Each propulsion system has its own set of advantages, challenges, and considerations, influencing the decision-making process for potential buyers.

As per Table 2.5,

1. Emissions: BEVs and PHEVs offer zero tailpipe emissions in the electric mode, making them environmentally friendly options compared to ICE vehicles. HEVs emit fewer emissions compared to conventional ICE vehicles, while still relying on fossil fuels for propulsion.
2. Range: BEVs are limited by battery capacity, resulting in shorter driving ranges compared to PHEVs, which offer a combination of electric range supplemented by the ICE. HEVs and ICE vehicles rely solely on fuel tanks for range, with the driving range dependent on fuel tank capacity and fuel efficiency.
3. Charging: BEVs require access to charging infrastructure for recharging the battery pack, while PHEVs require charging infrastructure

Table 2.5 Comparison of Propulsion Systems.

Aspect	*BEV*	*PHEV*	*HEV*	*ICE Vehicle*
Emissions	Zero tailpipe emissions	Zero tailpipe emissions in electric mode, reduced emissions in hybrid mode	Reduced emissions	Emissions from combustion of fossil fuels
Range	Limited by battery capacity	Electric range supplemented by ICE, longer total range compared to BEVs	Dependent on fuel tank capacity	Dependent on fuel tank capacity
Charging	Requires access to charging infrastructure	Requires access to charging infrastructure for the electric mode	No external charging required	Refueling at gas stations
Fuel efficiency	No fuel consumption	Variable fuel consumption depending on the driving mode	Improved compared to ICE vehicles	Dependent on fuel efficiency
Maintenance	Fewer moving parts, lower maintenance costs	Moderate maintenance requirements	Moderate maintenance requirements	Routine maintenance required
Driving dynamics	Instant torque, smoother acceleration	Smooth transition between electric and hybrid modes	Enhanced driving dynamics	Variable depending on the vehicle model

for the electric mode. HEVs and ICE vehicles do not require external charging infrastructure, relying solely on refueling at gas stations.

4. Fuel Efficiency: BEVs consume no fuel and offer zero tailpipe emissions, while PHEVs offer variable fuel consumption depending on the driving mode. HEVs offer improved fuel efficiency compared to conventional ICE vehicles, reducing fuel consumption and greenhouse gas emissions.
5. Maintenance: BEVs have fewer moving parts compared to ICE vehicles, resulting in lower maintenance costs over the vehicle's lifetime. PHEVs and HEVs require moderate maintenance, while ICE vehicles have routine maintenance requirements associated with ICEs.

6. Driving Dynamics: BEVs offer instant torque and smoother acceleration compared to ICE vehicles, providing a more responsive driving experience. PHEVs offer smooth transitions between electric and hybrid modes, while HEVs offer enhanced driving dynamics compared to conventional ICE vehicles.

By comparing various propulsion systems based on emissions, range, charging, fuel efficiency, maintenance, and driving dynamics, consumers can make informed decisions when selecting the most suitable electric vehicle for their needs. Each propulsion system offers unique advantages and considerations, catering to different driving preferences and environmental concerns.

Chapter 3

Electric Vehicle Components

3.1 ELECTRIC MOTORS AND DRIVE TRAINS

Electric motors and drivetrains are central components of electric vehicles (EVs), facilitating the conversion of electrical energy into mechanical energy for vehicle propulsion. Their design, efficiency, and integration play a crucial role in determining the performance, range, and reliability of EVs. This chapter explores the types of electric motors commonly used in EVs, their operational principles, and the configurations of drivetrains that complement them. Additionally, the section discusses advancements, challenges, and trends in motor and drivetrain technology.

Overview of Electric Motors in EVs

Electric motors are responsible for converting electrical energy from the battery pack into mechanical energy that drives the wheels. Unlike internal combustion engines (ICEs), electric motors provide instant torque [1–3], high efficiency, and minimal maintenance requirements. Table 3.1 provides an overview of the key types of electric motors used in EVs, highlighting their advantages and challenges.

Each motor type has its niche application in EVs, depending on factors such as cost, performance requirements, and availability of materials. For example, Tesla's earlier models relied heavily on induction motors, while its more recent vehicles have transitioned to permanent magnet synchronous motors (PMSMs) for improved efficiency. The primary configurations and internal components of EV powertrain sets are illustrated in Figure 3.1.

Key Components of Electric Motors

Electric motors in EVs consist of several critical components that contribute to their performance:

- Rotor and Stator: The rotor is the moving part, while the stator remains stationary. Together, they facilitate the generation of rotational motion.

 DOI: 10.1201/9781003608097-3

- Windings: These are copper or aluminum wires wound around the stator to create a magnetic field.
- Controller: This component regulates the electrical input to the motor, managing speed and torque.
- Cooling Systems: Essential for maintaining optimal operating temperatures, especially during high-load conditions.

Table 3.1 Key Types of Electric Motors in EVs.

Type of Motor	*Description*	*Advantages*	*Challenges*
Brushless DC motors (BLDC)	Operate without brushes, using electronic commutation	High efficiency, low maintenance, compact design	Higher cost due to sophisticated control systems
Induction motors	Use electromagnetic induction for operation, requiring no permanent magnets	Robust and cost-effective	Lower efficiency at partial load
Permanent magnet synchronous motors	Use permanent magnets for field excitation	High power density, superior performance	Dependence on rare-earth materials
Switched reluctance motors	Operate based on the reluctance principle with a simple construction	Cost-effective, durable	Noise and vibration concerns

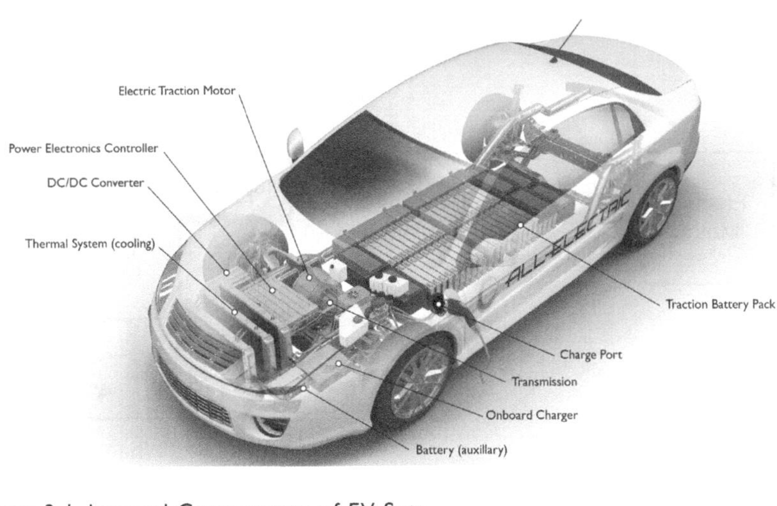

Figure 3.1 Internal Components of EV Sets.

Drivetrain Configurations in EVs

The drivetrain connects the motor to the wheels, translating mechanical energy into motion. EV drivetrains are distinct from traditional ICE drivetrains due to the simplicity of electric propulsion systems. Table 3.2 summarizes common drivetrain configurations in EVs.

Dual-motor all-wheel drive (AWD) configurations are particularly popular in high-performance EVs like Tesla's Model S Plaid and Rivian's R1T, offering exceptional acceleration and handling. On the other hand, single-motor configurations are common in budget-friendly models like the Nissan Leaf.

Table 3.2 Drivetrain Configurations in EVs.

Configuration	*Description*	*Advantages*	*Limitations*
Single-motor, single-axle	Single motor drives either the front or the rear axle	Simple design, cost-effective	Limited performance
Dual-motor, all-wheel drive	Two motors drive separate axles for improved traction and performance	Enhanced performance, better handling	Higher cost and complexity
In-wheel motors	Motors integrated directly into the wheels	High efficiency, compact design	Increased unsprung mass

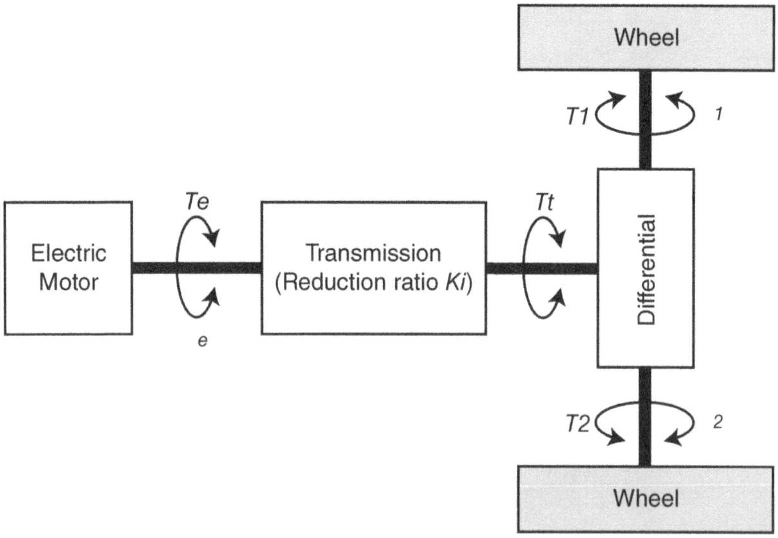

Figure 3.2. Drivetrain Configurations in EV Sets.

Operational Principles

Electric motors in EVs operate on the principles of electromagnetism. When electric current flows through the windings, it generates a magnetic field, causing the rotor to spin. The controller ensures smooth operation by managing voltage, current, and frequency.

Regenerative braking is an integral feature of EV motors, allowing the recovery of energy during braking. This energy [4–6] is fed back into the battery, enhancing overall efficiency and extending range.

Advancements in Electric Motors and Drivetrains

Several advancements are shaping the future of EV motors and drivetrains:

- High-Efficiency Motors: Innovations like solid-state inverters and improved materials are boosting motor efficiency.
- Integrated Drivetrains: Combining the motor, inverter, and transmission into a single unit reduces weight and increases efficiency.
- Advanced Cooling Techniques: Liquid cooling systems are becoming more prevalent, enabling motors to handle higher power densities.

Challenges and Solutions

Despite their advantages, electric motors and drivetrains face challenges:

- Material Dependence: PMSMs require rare-earth materials, which are expensive and environmentally taxing to extract.
- Thermal Management: High-performance motors generate significant heat, necessitating robust cooling solutions.
- Noise and Vibration: SRMs, while cost-effective, can produce undesirable noise and vibration.

To address these issues, researchers are exploring alternatives like ferrite magnets and novel motor topologies that reduce rare-earth dependence. Improved thermal materials and advanced soundproofing are also under development.

Future Trends

Future trends in electric motor and drivetrain technology focus on efficiency, sustainability, and integration:

- Wireless Charging Integration: Motors capable of receiving power wirelessly could revolutionize EV charging.
- Modular Drivetrains: Swappable modules for motors and drivetrains may simplify maintenance and upgrades.
- AI-Driven Optimization: Artificial intelligence is being used to optimize motor performance and energy management.

Electric motors and drivetrains are pivotal to the success of EVs, enabling efficient and reliable propulsion. As technology advances, these components are becoming more efficient, cost-effective, and sustainable. Continued innovation in motor designs, materials, and integration will play a critical role in the widespread adoption of EVs.

3.2 BATTERY PACKS AND CHEMISTRY

Battery packs are the cornerstone of EVs, directly influencing their range, performance, cost, and environmental impact. The design and chemistry [7–9] of battery packs are critical in determining the energy density, charging speed, and overall efficiency of an EV. This section delves into the structure of battery packs, explores various battery chemistries, and discusses the challenges and advancements in battery technology.

Overview of Battery Packs in EVs

A battery pack in an EV is more than just a power source: it is a sophisticated assembly of multiple interconnected cells designed to deliver consistent power while ensuring safety and reliability. The components of a typical battery pack include the following:

1. Cells: The fundamental units where energy is stored.
2. Modules: Groups of cells arranged to optimize energy output and thermal management.
3. Battery Management System (BMS): Monitors and regulates the battery's operation, including charge and discharge cycles, temperature, and safety parameters.
4. Thermal Management System: Maintains the pack's temperature within optimal limits to ensure performance and longevity.

Table 3.3 summarizes the key components of an EV battery pack and their respective roles.

Table 3.3 Key Components of an EV Battery Pack.

Component	*Description*	*Role in the Battery Pack*
Cells	Small energy storage units (cylindrical, prismatic, or pouch formats)	Store electrical energy in chemical form
Modules	Organized clusters of cells	Facilitate energy distribution and thermal stability
BMS	Integrated circuitry and software	Regulates charge/discharge, monitors health, ensures safety
Thermal management system	Cooling/heating systems using liquids, air, or phase-change materials	Prevents overheating and ensures consistent performance

The synergy between these components ensures the safe and efficient functioning of the battery pack.

Battery Chemistries in EVs

The chemistry of a battery determines its performance characteristics, such as energy density, lifespan, safety, and cost. The most common battery chemistries used in EVs include the following:

- Lithium-Ion (Li-ion): The dominant technology in EVs, known for high energy density and efficiency.
- Nickel–Metal Hydride (NiMH): Previously popular in hybrid vehicles, now largely replaced by Li-ion.
- Solid-State Batteries: Emerging technology promising higher energy density and safety.

Table 3.4 compares these battery chemistries, focusing on their key attributes.

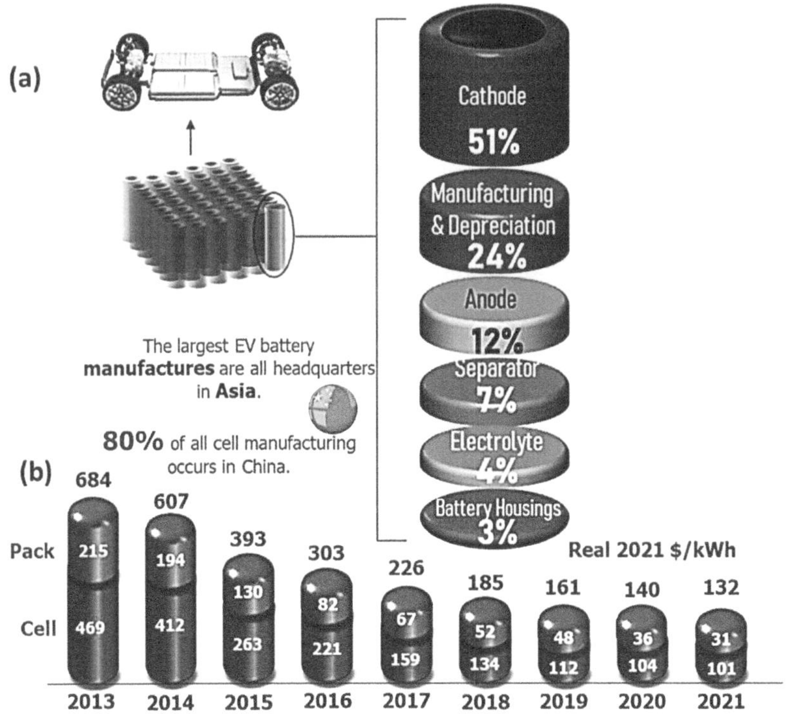

Figure 3.3 Battery Chemistry Levels.

Table 3.4 Comparison of Battery Chemistries in EVs.

Chemistry	*Energy Density (Wh/kg)*	*Lifespan (Charge Cycles)*	*Cost ($/kWh)*	*Safety*	*Applications*
Li-ion	150–250	1,000–3,000	100–150	Moderate	Dominant in EVs due to high performance
NiMH	60–120	500–1,000	200–300	High	Used in older hybrids; now declining
Solid-state	300–500	>5,000	300+	Very high	Experimental; future EV applications

Li-ion: These are the standard for EVs due to their high energy density, efficiency, and relatively lower cost compared to other advanced chemistries. However, they require careful thermal management to prevent overheating and thermal runaway.

NiMH: Known for durability and safety, NiMH batteries are bulky and less efficient, leading to their declining use in modern EVs.

Solid-State Batteries: Offering significant improvements in energy density and safety, solid-state batteries are currently in the research and development stage but hold the potential to revolutionize EV technology.

Battery Performance Metrics

Battery performance is evaluated based on several key metrics [10–12]:

- Energy Density: Determines how much energy the battery can store per unit of weight or volume.
- Cycle Life: The number of charge/discharge cycles the battery can endure before its capacity degrades significantly.
- Charging Speed: Time required to recharge the battery to a specified capacity.
- Safety: Resistance to issues such as overheating, leakage, and short-circuiting.

These metrics influence the overall performance and user experience of EVs. For instance, batteries with higher energy densities enable longer ranges, while faster charging speeds improve convenience for users.

Advancements in Battery Chemistry

The EV industry is witnessing rapid advancements in battery chemistry aimed at overcoming current limitations:

- Silicon Anodes: Replace graphite anodes in Li-ion batteries to improve energy density.
- Cobalt-Free Chemistries: Reduce reliance on cobalt for cost reduction and ethical sourcing.
- Recycling Technologies: Address environmental concerns by enabling efficient recovery of valuable materials from used batteries.

Challenges and Solutions in Battery Technology

Despite significant progress, battery technology faces several challenges:

- Cost: Battery packs constitute a significant portion of an EV's cost. Efforts are underway to reduce costs below $100/kWh to make EVs more affordable.
- Resource Scarcity: Materials like lithium and cobalt are limited and have ethical sourcing concerns.
- Recycling and Disposal: Ensuring sustainable end-of-life management for batteries is a critical issue.

To address these challenges, researchers are developing alternative materials, enhancing recycling processes, and exploring second-life applications for retired EV batteries.

Case Study: Tesla's Battery Innovation

Tesla has been a leader in advancing battery technology. Its use of cylindrical Li-ion cells with nickel–cobalt–aluminum (NCA) chemistry has set industry standards. The introduction of tableless electrodes in its 4,680 cells has significantly improved thermal management [13–15] and reduced manufacturing costs.

Future Outlook

Future trends in EV battery technology include the following:

- Solid-State Batteries: Expected to become mainstream in the next decade due to their superior performance and safety.
- Battery-as-a-Service (BaaS): Models where consumers lease batteries, reducing the upfront cost of EVs.
- Wireless Charging Integration: Batteries capable of receiving power wirelessly could transform EV charging infrastructure.

Battery packs and their chemistries are pivotal to the success of EVs. The ongoing innovations in materials, design, and manufacturing processes are driving improvements in performance, cost, and sustainability. As the industry continues to evolve, the advancements in battery technology will play a crucial role in shaping the future of electric mobility.

3.3 POWER ELECTRONICS AND INVERTERS

Power electronics and inverters are indispensable components in EVs, responsible for managing energy flow and converting electrical energy into usable forms for vehicle operation. These systems ensure efficient operation of the electric motor and other subsystems, contributing to the overall performance, range, and reliability of EVs. This section explores the role of power electronics and inverters; their key components, types, and advancements; and their challenges and solutions.

Overview of Power Electronics in EVs

Power electronics in EVs manage the conversion and control of electrical power from the battery to various components, such as the motor, auxiliary systems, and charging systems. Key tasks include voltage regulation, current control, and ensuring high efficiency during energy transfer. Table 3.5 highlights the main functions of power electronics in EVs.

Power electronics are essential for seamless energy management in EVs, improving both performance and energy efficiency.

Role of Inverters in EVs

Inverters are a subset of power electronics that specifically handle the conversion of direct current (DC) from the battery into alternating current (AC)

Table 3.5 Functions of Power Electronics in EVs.

Function	*Description*
DC–DC conversion	Converts high-voltage DC from the battery to lower voltages for auxiliary systems
DC–AC inversion	Converts DC power from the battery to AC power for the electric motor
Regenerative braking	Manages the conversion of mechanical energy into electrical energy during braking
Battery management	Regulates charging and discharging to ensure battery safety and longevity

for the motor. This conversion is critical because most high-performance electric motors, such as induction and permanent magnet synchronous motors, operate on AC.

Key features of inverters in EVs include the following:

- Pulse Width Modulation (PWM): Controls the output voltage and frequency to regulate motor speed and torque.
- Thermal Management: Ensures the inverter operates within safe temperature limits to maintain efficiency.
- Regenerative Functionality: Supports energy recovery during braking.

Table 3.6 compares the types of inverters commonly used in EVs.

Among these, voltage source inverter (VSI) is the most widely used type due to its balance of simplicity and performance.

Key Components of Power Electronics and Inverters

Power electronics and inverters consist of several critical components that work in unison to regulate energy flow:

1. Semiconductor Switches: These include insulated gate bipolar transistors (IGBTs) and metal-oxide-semiconductor field-effect transistors (MOSFETs). They control the switching of electrical power.
2. Capacitors and Inductors: These components manage voltage spikes and smoothen current flow.
3. Microcontrollers: Handle real-time control of power conversion processes.
4. Cooling Systems: Prevent overheating, which can degrade performance and reliability.

Table 3.6 Types of Inverters in EVs.

Type of Inverter	*Description*	*Advantages*	*Challenges*
Voltage source inverter	Converts DC voltage into AC with fixed amplitude	High efficiency, simple design	Limited to specific voltage ranges
Current source inverter	Converts DC current into AC for the motor	Robust, reliable	Bulky, lower efficiency
Multilevel inverter	Produces AC with multiple voltage levels	Lower harmonics, higher efficiency	Complex control, higher cost

Efficiency and Losses

Efficiency is a critical factor for power electronics and inverters in EVs, as losses can reduce range and overall performance. Losses typically occur due to the following:

- Conduction Losses: Resistance in semiconductors during current flow.
- Switching Losses: Energy lost during the on/off transitions of semiconductor switches.
- Thermal Losses: Heat generated due to inefficiencies.

Table 3.7 provides an overview of strategies to minimize these losses.

By adopting these strategies, manufacturers aim to enhance the energy efficiency of EVs.

Advancements in Power Electronics and Inverters

Advancements in materials and design are revolutionizing power electronics and inverters:

- Silicon Carbide (SiC) and Gallium Nitride (GaN): These wide-bandgap materials offer superior performance, enabling higher efficiency and compact designs.
- Integrated Systems: Combining inverters with motors and chargers into a single unit reduces weight and enhances efficiency.
- Smart Control Algorithms: Advanced algorithms improve the responsiveness and adaptability of inverters to dynamic driving conditions.

Challenges and Solutions

Power electronics and inverters face several challenges:

- Thermal Management: High-power applications generate significant heat, requiring advanced cooling solutions.

Table 3.7 Strategies to Minimize Losses in Power Electronics.

Type of Loss	*Strategy*
Conduction losses	Use of low-resistance semiconductor materials like silicon carbide (SiC)
Switching losses	Implementation of soft-switching techniques and advanced gate drivers
Thermal losses	Deployment of efficient cooling systems such as liquid cooling or phase-change materials

- Cost: Wide-bandgap materials like SiC and GaN are expensive, increasing manufacturing costs.
- Reliability: Ensuring consistent performance under varying operating conditions is critical.

To address these challenges, research is focusing on developing cost-effective materials, improving thermal management techniques, and enhancing durability through robust design practices.

Future Trends

Future trends in power electronics and inverters include the following:

- Wireless Power Transfer: Development of wireless charging systems integrated with power electronics.
- Bidirectional Inverters: Enabling vehicle-to-grid (V2G) applications for energy feedback into the grid.
- AI-Driven Optimization: Leveraging artificial intelligence for predictive maintenance and real-time energy management.

Power electronics and inverters are fundamental to the functionality and efficiency of EVs. Advances in materials, design, and control systems are driving improvements in energy conversion and management, contributing to the widespread adoption of EVs. Continued innovation in these areas is essential for the development of next-generation EVs that are more efficient, cost-effective, and sustainable.

3.4 REGENERATIVE BRAKING SYSTEMS

Regenerative braking systems (RBSs) are a defining feature of EVs, enabling energy recovery during braking. By converting kinetic energy into electrical energy, the RBS enhances energy efficiency, extends the driving range, and reduces wear on mechanical brakes. This section delves into the principles, components, types, advantages, challenges, and advancements in RBSs, supplemented by relevant data in tabular form.

Principles of Regenerative Braking

Regenerative braking works on the principle of energy conversion. When an EV slows down, the electric motor operates in reverse, acting as a generator to convert kinetic energy into electrical energy, which is stored back in the battery. Unlike traditional friction braking, which dissipates energy as heat, regenerative braking reclaims a portion of the energy, improving the overall efficiency of the vehicle.

Table 3.8 provides a comparison between traditional braking systems and RBSs.

This comparison underscores the efficiency and sustainability benefits of regenerative braking in EVs.

Key Components of Regenerative Braking Systems

A regenerative braking system in EVs comprises several components working in harmony:

1. Electric Motor/Generator: The motor switches to generator mode during braking to produce electrical energy.
2. Battery Pack: Stores the energy recovered during braking for future use.
3. Power Electronics: Regulates the conversion and flow of energy between the motor and battery.
4. Brake Control Unit (BCU): Coordinates between regenerative braking and traditional friction braking to optimize performance and safety.

Types of Regenerative Braking Systems

RBSs can be classified into three main types based on their integration and operation:

1. Series Regenerative Braking: Prioritizes regenerative braking, with mechanical brakes serving as a backup.
2. Parallel Regenerative Braking: Simultaneously uses both regenerative and mechanical brakes.
3. Blended Braking: Seamlessly integrates regenerative and mechanical braking for smooth operation.

Table 3.9 highlights the characteristics of these systems.

Table 3.8 Comparison of Traditional Braking Systems and RBSs.

Feature	*Traditional Braking System*	*RBS*
Energy conversion	Converts kinetic energy into heat	Converts kinetic energy into electrical energy
Energy recovery	None	Partial recovery
Brake wear	High	Low
Environmental impact	Generates heat and particulate matter	Reduces energy wastage, environmentally friendly
System complexity	Simple	Requires integration with electric motors and BMS

Table 3.9 Types of RBSs.

Type	*Description*	*Advantages*	*Challenges*
Series	Regenerative braking takes precedence; mechanical brakes engage only when needed	High energy recovery potential	Limited mechanical braking response
Parallel	Both systems work concurrently	Enhanced braking power	Increased system complexity
Blended	Control system balances regenerative and mechanical braking	Smooth braking experience, safety focused	Requires advanced coordination algorithms

Blended braking systems are widely adopted in modern EVs due to their balance of energy recovery and user experience.

Advantages of Regenerative Braking Systems

RBSs offer several advantages:

- Energy Efficiency: RBS increases the overall energy efficiency of EVs by reclaiming energy that would otherwise be wasted.
- Extended Range: By feeding recovered energy back into the battery, RBS helps extend the driving range.
- Reduced Brake Wear: RBS minimizes reliance on friction brakes, leading to lower maintenance costs and longer brake life.
- Environmental Benefits: By reducing energy wastage and brake wear, RBS contributes to reduced emissions and particulate matter.

Challenges in Regenerative Braking Systems

Despite its benefits, RBSs face certain challenges:

- Energy Recovery Limits: The amount of energy recovered depends on factors like speed and braking force.
- System Complexity: Integration with the motor, battery, and braking system increases design and manufacturing complexity.
- Driver Perception: The braking feel in RBSs can differ from traditional systems, requiring adjustments for user comfort.

Table 3.10 provides a summary of these challenges and possible solutions.

These solutions are key to enhancing the efficiency and acceptance of RBSs.

Table 3.10 Challenges and Solutions in RBSs.

Challenge	*Description*	*Potential Solutions*
Energy recovery limits	Energy recovery is less effective at low speeds	Improve motor efficiency and optimize control algorithms
System complexity	Increased integration requirements with other EV components	Use modular designs and advanced software controls
Driver perception	Braking behavior may feel unfamiliar to drivers	Develop adaptive braking systems with customizable settings

Advancements in Regenerative Braking Technology

Continuous advancements are being made in regenerative braking technology to overcome existing limitations and improve efficiency:

- Integration with Advanced Driver-Assistance Systems (ADAS): Enhances coordination between braking and safety systems.
- Wireless Energy Transfer: Research into wireless systems aims to directly feed recovered energy into grid infrastructure.
- Smart Control Algorithms: Algorithms optimize the distribution of braking forces between regenerative and mechanical systems for improved energy recovery.

Case Study: Tesla's Regenerative Braking System

Tesla has pioneered advanced RBSs, incorporating adaptive controls to optimize energy recovery based on driving conditions. The company's use of blended braking ensures smooth transitions between regenerative and mechanical braking, enhancing both efficiency and user experience.

Future Trends

Future developments in RBSs focus on increasing energy recovery and user convenience:

- Bidirectional Charging: Allows energy recovered during braking to be used for V2G applications.
- Enhanced Predictive Controls: Use Artificial Intelligence (AI) to predict braking needs and maximize energy recovery.
- High-Efficiency Materials: Employ lightweight and durable materials to improve system performance.

RBSs are a cornerstone of EV efficiency and sustainability. By reclaiming energy during braking, these systems enhance the range, reduce maintenance costs, and contribute to environmental benefits. Despite challenges, advancements in technology and control systems continue to drive improvements in regenerative braking, ensuring its pivotal role in the future of electric mobility.

3.5 VEHICLE CONTROL SYSTEMS

Vehicle control systems (VCSs) are the brains of modern EVs, orchestrating the operation of critical components such as the motor, battery, power electronics, and braking systems. These systems ensure optimal performance, safety, and energy efficiency through advanced algorithms and real-time monitoring. This section explores the architecture, functions, types, and advancements in VCSs, along with challenges and solutions.

Overview of Vehicle Control Systems

VCSs in EVs manage and coordinate the interaction between different subsystems. By collecting data from sensors and executing predefined algorithms, a VCS optimizes vehicle performance under varying conditions. Table 3.11 highlights the primary functions of VCSs in EVs.

These functions enable seamless operation and efficient energy use in EVs.

Architecture of Vehicle Control Systems

A typical VCS consists of hardware and software components that work together to manage the vehicle's operations. The key elements include the following:

- Central Control Unit (CCU): The primary processor that executes control algorithms.
- Sensors: Devices that collect real-time data such as speed, temperature, battery state, and motor performance.

Table 3.11 Primary Functions of VCSs in EVs.

Function	*Description*
Power management	Controls power distribution among the battery, motor, and auxiliary systems
Thermal management	Regulates the temperature of components to ensure efficiency and safety
Drive control	Manages motor speed and torque based on driver inputs and road conditions
Regenerative braking	Coordinates energy recovery during braking to enhance efficiency
Fault detection	Monitors the system for errors and ensures safe operation through protective measures

- Actuators: Components that convert control signals into physical actions, such as adjusting motor torque or activating brakes.
- Communication Network: A system like a controller area network (CAN) or Ethernet that facilitates data exchange between subsystems.

Types of Vehicle Control Systems

VCSs in EVs can be categorized based on their functionality and integration level. Table 3.12 compares the major types of VCSs.

Integrated VCSs are commonly used in modern EVs, while autonomous VCSs are the focus of ongoing research and development for future applications.

Key Features of Vehicle Control Systems

The effectiveness of a VCS depends on its ability to adapt to varying operational scenarios and driver inputs. Some key features include the following:

1. Adaptive Control: Adjusts to changes in road conditions, load, and driver behavior.
2. Predictive Algorithms: Anticipates system needs based on historical data and real-time inputs.
3. Energy Optimization: Minimizes energy consumption through intelligent power distribution and regenerative braking.
4. User Customization: Allows drivers to select driving modes, such as eco, sport, or comfort.

Table 3.12 Types of VCSs.

Type	*Description*	*Advantages*	*Challenges*
Basic VCS	Manages essential operations like motor control and battery management	Simple and cost-effective	Limited features and adaptability
Integrated VCS	Combines control of multiple subsystems for better efficiency and coordination	Enhanced performance, lower energy loss	Increased complexity and cost
Autonomous VCS	Includes advanced features like ADAS and autonomous driving controls	High functionality and safety integration	Requires significant computational power

Challenges in Vehicle Control Systems

Despite their advanced capabilities, VCSs face several challenges:

- System Complexity: The integration of multiple subsystems requires sophisticated algorithms and communication protocols.
- Reliability: Ensuring fault tolerance and redundancy is crucial for safety-critical operations.
- Cost: Developing and deploying advanced VCSs increases the overall cost of EVs.

Table 3.13 outlines these challenges and potential solutions.

Advancements in Vehicle Control Systems

Advancements in technology are driving the evolution of VCSs:

1. AI: AI algorithms improve decision-making and predictive capabilities, particularly for autonomous VCSs.
2. Over-the-Air (OTA) Updates: Enable real-time software improvements without requiring physical servicing.
3. Edge Computing: Reduces latency by processing data closer to the source, improving system responsiveness.
4. Integration with the Internet of Things (IoT): Enhances connectivity for features like remote diagnostics and fleet management.

Case Study: Tesla's Vehicle Control System

Tesla's VCS exemplifies state-of-the-art control systems in EVs. The company integrates features such as the following:

- Autopilot: Semi-autonomous driving capabilities with lane-keeping and adaptive cruise control.

Table 3.13 Challenges and Solutions in VCSs.

Challenge	*Description*	*Potential Solutions*
System complexity	Integration of numerous subsystems and control algorithms	Modular architecture, use of standardized protocols
Reliability	High reliance on sensors and software for safety-critical tasks	Implementation of fault-tolerant systems and redundancy
Cost	High development and implementation costs	Use of open-source platforms and scalable solutions

- Energy Management: Intelligent power allocation between dual motors in AWD models.
- User-Centric Features: OTA updates and customizable driving modes.

Tesla's VCS demonstrates the potential of advanced control systems in enhancing both performance and user experiences.

Future Trends

The future of VCSs is oriented toward higher levels of automation, connectivity, and efficiency:

- Level 4/5 Autonomy: Full self-driving capabilities with minimal human intervention.
- Blockchain Integration: Ensures secure data exchange and enhances cybersecurity.
- Cloud-Based Control: Centralized control systems leveraging cloud computing for enhanced scalability and real-time analytics.

VCSs are pivotal in optimizing the performance, safety, and efficiency of EVs. Through advanced algorithms and seamless integration of subsystems, VCSs enable intelligent energy management and enhanced user experiences. As technology evolves, future VCSs will continue to push the boundaries of automation and connectivity, playing a key role in the next generation of electric mobility.

REFERENCES

1. Acar, E., Jain, N., Ramu, P. *et al.* A survey on design optimization of battery electric vehicle components, systems, and management. *Struct Multidisc Optim* 67, 27 (2024). https://doi.org/10.1007/s00158-024-03737-7
2. Suh, K., & Moon, J. Electric vehicle architecture design based on database. *Int J Automot Technol* 25, 427–444 (2024). https://doi.org/10.1007/s12239-024-00035-5
3. Shewalkar, A.G., Dhoble, A.S., & Thawkar, V.P. Review on cooling techniques and analysis methods of an electric vehicle motor. *J Therm Anal Calorim* 149, 5919–5947 (2024). https://doi.org/10.1007/s10973-024-13091-x
4. An, X., Liu, Y., Liu, Z. *et al.* Research on rapid cycle test method in low-temperature of electric vehicle driving range based on China light-duty vehicle test cycle. *Int J Automot Technol* (2024). https://doi.org/10.1007/s12239-024-00145-0
5. Velasquez, J.D.O., Moreno, J.A.G., & Aldana, N.L.D. Experimental platform for studying energy regeneration in electric vehicle powertrains. *J Power Electron* 24, 1751–1765 (2024). https://doi.org/10.1007/s43236-024-00836-y

6. Wang, G., Wang, L., Wang, L. *et al.* Electric vehicle charging load modeling based on influence factor analysis. *J Shanghai Jiaotong Univ (Sci)* (2023). https://doi.org/10.1007/s12204-023-2663-2
7. Gao, C., Zhao, X., Zhu, W. *et al.* Impact load between spherical obstacles and electric vehicle battery packs. *J Mech Sci Technol* 38, 2845–2854 (2024). https://doi.org/10.1007/s12206-024-0504-3
8. Abdullah, M., Moin, L., Ahmed, F. *et al.* Electric vehicle battery temperature control using fuzzy logic. *Aut Control Comp Sci* 58, 237–251 (2024). https://doi.org/10.3103/S0146411624700135
9. Tan, D., Pang, Z., Li, M. *et al.* Investigation of motor bearing thermal characteristics for in-wheel electric vehicle applications. *J Mech Sci Technol* 38, 6939–6952 (2024). https://doi.org/10.1007/s12206-024-1144-3
10. Pan, L., Breton, F., & Fourmann, J. development of a new Al-Fe-Ni alloy for electric vehicle applications. *Inter Metalcast* 18, 2841–2845 (2024). https://doi.org/10.1007/s40962-024-01307-7
11. Ge, S., Yan, J., Yang, Y. *et al.* Multifield coupled dynamics modeling and vibration characteristics of electric-vehicle electric drive systems. *Nonlinear Dyn* 112, 19669–19689 (2024). https://doi.org/10.1007/s11071-024-10055-6
12. Cheng, A.L., Fuchs, E.R.H., & Michalek, J.J. US industrial policy may reduce electric vehicle battery supply chain vulnerabilities and influence technology choice. *Nat Energy* 9, 1561–1570 (2024). https://doi.org/10.1038/s41560-024-01649-w
13. Rao, J.V.G.R., & Venkateshwarlu, S. Soft-switching dual active bridge converter-based bidirectional on-board charger for electric vehicles under vehicle-to-grid and grid-to-vehicle control optimization. *J Eng Appl Sci* 71, 49 (2024). https://doi.org/10.1186/s44147-024-00384-z
14. Agbesi, P.K., Ruffino, R., & Hakovirta, M. Creating an optimal electric vehicle ecosystem: An investigation of electric vehicle stakeholders and ecosystem trends in the US. *SN Bus Econ* 4, 28 (2024). https://doi.org/10.1007/s43546-024-00624-7
15. Saravanan, K., Sivasubramanian, M., Gopinath, N.P. *et al.* A 31 L multilevel inverter topology with less switching devices for hybrid electric vehicle applications. *Sci Rep* 14, 27459 (2024). https://doi.org/10.1038/s41598-024-78529-6

Chapter 4

Battery Technology

4.1 LITHIUM-ION BATTERIES

The lithium (Li) ion has proven to be a keystone technology in the case of modern electric vehicles (EVs). High energy density, long cycle life, and decreasing cost have made Li ion the flagship technology. The Li-ion battery development has helped in building EV technology because they offered reliable and efficient energy storage solutions. In this chapter, basic principles, material compositions, performance characteristics, and challenges associated with the Li-ion battery are elaborated while underlining its position in the EV ecosystem.

A Li ion basically works on the electrochemical reaction between Li ions and materials available in electrodes. Generally, a Li-ion battery has three main components: one is called the positive electrode-anode being referred to as the negative electrode, and the substance that allows the movement of ions is called an electrolyte. In the charging process of a Li-ion battery, Li ions move out of the cathode and pass to the anode material, where they intercalate into the material through the electrolyte. During the discharge, this is reversed, and an electric current that advances the EV is delivered to the motor. Li-ion batteries have the energy density between 150 and 200 Wh/kg. This makes them optimal for use in EVs due to the weight and space occupied by the device for energy storage.

A cathodic material has considerable influence on the performance and energy density of a battery and its safety. The most commonly used cathodes are lithium cobalt oxide (LCO), lithium manganese oxide (LMO), lithium nickel manganese cobalt oxide (NMC), and lithium iron phosphate (LFP). Among those, NMC cathode is the most popular in EV applications because of its balanced energy density, high safety, and lesser cost. So, a NMC cathode reaches a value of around 160–200 Wh/kg energy density, which is fairly good enough for EVs for longer ranges. In general, the anode material would be graphite that possesses high potential for Li-ion storage and structural stability.

The electrolyte of a Li-ion battery is made of Li salts dissolved in organic solvents. This means it ensures good ionic conductivity over a wide stable

DOI: 10.1201/9781003608097-4

window. But organic solvents are flammable and hence pose safety concerns for such cells; therefore, high-thermal-management requirements coupled with enhanced safety procedures for thermal events are always associated with such an EV application. This means that the separator, the porous polymer (POE) film in this case, creates an electrochemical effect between electrodes without any form of physical separation, allowing the passage of ions. This highly advanced separator technology is very safe, in the sense that the risks of short circuits and the possibility of thermal runaway are reduced.

Good performance characteristics indicate good performances for Li-ion batteries in the EV application. This translates [1–3] to a high charge–discharge efficiency, frequently achieving results greater than 95%. This efficiency minimizes the loss of energy during charging and releasing processes; hence, it is highly efficient. Li-ion battery systems also have an optimal operating temperature range. Temperatures may vary from –20 to 60°C, although any temperature outside of that would likely decrease the performance or life of a battery. This also counts for the cycle life of batteries, which ranges from 1,000 to 3,000 cycles before some significant capacity is lost, and an example for a perspective is where an EV with a Li-ion battery pack designed to withstand up to 1,500 cycles can travel an average of 200 km per charge over around 300,000 km.

This was driven by improvements in manufacturing processes, economies of scale, and material optimization. Further reductions are needed to bring costs into the realm of a mass market. Table 4.1 presents the Li-ion battery cost trajectory over the past decade and its impact on vehicle penetration levels.

Table 4.1 illustrates the relationship between the decrease in cost of the batteries and the increase in EV penetration. The more expensive they are, the less the EVs compete with the internal combustion vehicles (ICVs); thus, penetration is not promoted.

There are also significant issues of reliance on critical raw materials, such as Li, cobalt, and nickel. The extraction and processing of these materials could seriously impact the environment and also raise ethical concerns. Cobalt mining is especially associated with labor exploitation and environmental degradation, particularly from operations in the Democratic

Table 4.1 Cost Trajectory of Li-Ion Batteries (2010–2022).

Year	*Cost per kWh (USD)*	*Cumulative EV Sales (Million Units)*
2010	1,200	0.2
2015	500	1.5
2020	200	10
2022	150	20

Republic of Congo, which accounts for over 60% of cobalt globally [4–6]. To reduce the cobalt content, research on low cobalt and cobalt-free cathodes has grown, such as NMC 811, where the ratio of nickel to manganese to cobalt is 8:1:1, and LFP. Table 4.2 shows some of the important properties of the more popular cathode materials used in Li-ion batteries.

Table 4.2 shows that LFP batteries have safety over the others due to more extended cycle life and have lower cost; therefore, such batteries can be used for applications whose requirements are of long lifetime and low cost. Energy density is relatively less when compared to NMCs; so it cannot be used for long-range EVs.

Due to recycling and end-of-life management, the Li-ion environmental footprint will be significantly reduced. These technologies recover the valuable materials present, such as Li, cobalt, and nickel, besides reducing virgin resource demand. Pyrometallurgical and hydrometallurgical processes are

Table 4.2 Key Properties of Common Cathode Materials.

Cathode Material	*Energy Density (Wh/kg)*	*Cost (USD/kg)*	*Safety*	*Cycle Life (Cycles)*
LCO	140–160	35	Moderate	500–1,000
NMC	160–200	25	High	1,000–2,000
LFP	140–170	15	Very high	2,000–3,000
LMO	120–140	20	High	500–1,500

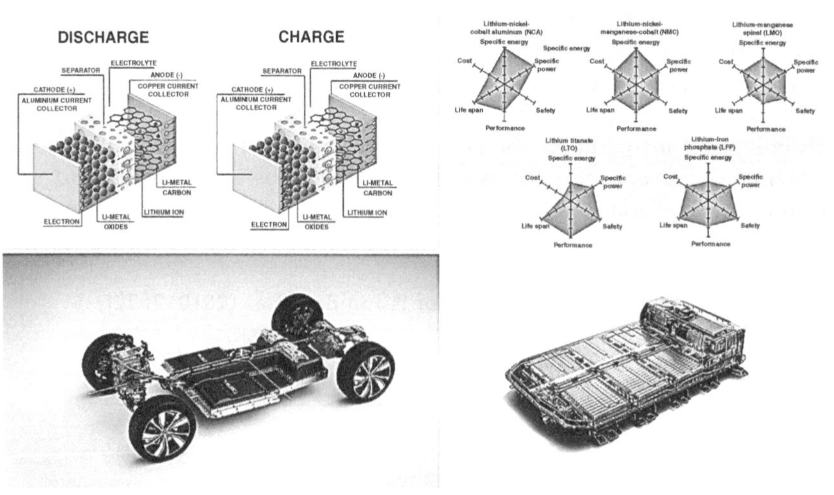

Figure 4.1 Li-Ion Battery Configurations.

generally considered for battery recycling; these have pros and cons to their operations. As a result, the global Li-ion battery recycling rate stands at only around 5–10%, and thus still holds tremendous opportunities for improvement. Within that backdrop, cost-effective recycling is crucial for achieving sustainable growth for the EV industry.

In a nutshell, Li-ion batteries represent the innovation technology for EVs. Along with energy density, efficiency, and cycle life [7–9], no equal capability is exhibited by such batteries. Improvement in material science, manufacturing, and recycling continues to raise the bar for Li-ion batteries with respect to their performance and sustainability. However, all the disadvantages related to cost and environment of raw materials supply still confront further innovations in this sphere. In case of persistent growth of EV demands, Li-ion batteries will stand at the nucleus of sustainable systems of transportations.

4.2 NICKEL–METAL HYDRIDE BATTERIES

The most applicable energy storage solution in hybrid electric vehicles (HEVs) and other applications that demand longevity at moderate energy density is that provided by nickel–metal hydride (NiMH) batteries. NiMH batteries are known for their durability, offering a high cycle count with environmental friendliness. There is certainly a very strong linkage between nickel–cadmium (NiCd), both on its older side and on the newer side, and Li ion with regard to properties related to it to make it as one of the best candidates for energy storage in hybrid EVs. This describes the fundamental principles, properties, characteristics, performance metrics, issues, and applicability of NiMH batteries to EVs.

A NiMH battery works on reversible electrochemical reactions with hydrogen ions or protons. The main parts include a positive electrode or cathode made of nickel oxyhydroxide (NiOOH); a negative electrode or anode, which is a hydrogen-absorbing alloy; and an electrolyte, which is aqueous potassium hydroxide (KOH). Hydrogen ions move from the anode to the cathode, creating electrons and electrical current when the battery is in the process of discharging. In charging, the process is reversed, which revives the capacity of the battery. This reversibility with a very robust electrode design makes NiMH batteries for demanding applications.

NiMH typically has an energy density within the 60–120 watt-hour per kilogram range, that is, less than that of Li ions but more than that of nickel–cadmium. A moderate energy density has made NiMH batteries historically preferential for HEV applications, where the demand for energy storage is not as demanding as in fully EV applications. NiMH batteries achieve a cycle life greater than 2,000 cycles or more depending on Depth of Discharge (DoD) and operating conditions, which is an important advantage in hybrid applications that are seeing frequent charge–discharge cycles.

NiMH also exhibits excellent thermal stability at a range of −20–70°C. It is less susceptible to thermal runaway than Li-ion cells, which makes it much safer as a cell. That it is made of non-toxic and recyclable material makes this battery much more environmentally friendly. Table 4.3 presents the comparative key performance metrics between NiMH and other chemistries.

Table 4.3 shows that NiMH batteries have a balance among energy density, cycle life, and cost. Their thermal stability and durability make them best suited for hybrid EVs when reliability is more important than high energy density.

One important characteristic of NiMH batteries is tolerance for overcharging and deep discharging without significant degradation. This is one of the properties of the internal oxygen recombination mechanism of the battery, which does not allow pressure buildup due to overcharging. Oxygen formed at the cathode combines with hydrogen at the anode, thus ensuring chemical equilibrium and safe operation. They self-discharge faster, however. For example, NiMH loses 20–30% of its charge per month if unused. This self-discharge rate is significantly higher compared to Li-ion batteries, which have to be recharged more often in order to sustain their performance.

The most commonly used anode material for NiMH is a hydrogen-absorbing alloy made of rare-earth elements, nickel, and other metallic constituents like cobalt, manganese, or aluminum. The selection of various alloys causes variations in the hydrogen storage capacity of the battery, charging efficiency, and cycle life. Among them, very often, AB5- and AB2-type alloys are used: AB5-type alloys lanthanum–nickel–cobalt, and AB2-type alloys titanium–zirconium–nickel. While AB5-type alloys offer superior stability and cycling life [10–12], AB2-type alloys present more energy density but require strict processing techniques. Properties of different alloy types are presented in Table 4.4.

As shown in Table 4.4, AB5 alloys are favored for their stability and long cycle life; therefore, they are used extensively in commercial NiMH batteries. However, AB2 alloys have received more attention lately, especially when high energy densities are required, such as in advanced hybrid systems.

Table 4.3 Comparison of Battery Chemistries.

Parameter	*NiMH*	*Li Ion*	*NiCd*
Energy density (Wh/kg)	60–120	150–200	45–80
Cycle life (cycles)	1,000–2,000	1,000–3,000	500–1,500
Operational temperature range	−20 to 70°C	−20 to 60°C	−20 to 50°C
Cost per kWh (USD)	200–300	150–250	300–500

Table 4.4 Properties of Hydrogen-Absorbing Alloys.

Alloy Type	*Hydrogen Storage Capacity (wt%)*	*Stability*	*Cycle Life*	*Manufacturing Complexity*
AB5 (LaNi5Co)	1.4–1.6	High	>2,000	Moderate
AB2 (TiZrNi)	1.8–2.0	Moderate	1,500–2,000	High

Environmental and cost considerations of NiMH batteries also need to be discussed. Unlike the critical materials Li and cobalt used in Li-ion batteries, NiMH is composed of abundant and recyclable elements. This dramatically reduces the environmental footprint of this battery and contributes to sustainability. However, the use of rare-earth elements in the anode alloy brings with it a host of issues related to mining practice and the reliability of the supply chain. For example, more than 85% of rare earth is found in China, and consequently, there is geopolitical and economic dependence. Minimizing dependence is through work on the development of rare-earth-free alloys or other alternative anode materials.

Li-ion batteries are challenging NiMH-based batteries in the EV market. The latter ones provide better energy density, lesser weight, and faster charging times and are thus used for fully EVs. However, in HEVs, the NiMH still finds applicability because in such cases durability and safety along with low cost are preferred instead of the requirement of high energy density. Applications of NiMH batteries across different domains are presented in Table 4.5.

As indicated in Table 4.5, high charge–discharge cycle frequencies and thermal stability make them suitable for HEVs. For consumer electronics, moderate energy density coupled with environmental safety makes them the perfect candidate for rechargeable cameras, handheld tools, etc. Reliability and tolerance to overcharge make them the ideal candidate for backup power systems in critical infrastructure.

In summary, NiMH is a robust energy storage battery proven at the test track in the HEV and similar applications: it shows average energy density, long lifetimes, and thermal stability, thus its use in situations where higher demands for strength and safety are anticipated. It suffers, however, from self-discharge rates, material costs, and competition by Li-ion batteries. This will certainly promise further development of electrode material and manufacturing techniques to attain better performance and sustainability for the NiMH batteries, and hence secure their place in the yet-evolving energy storage landscape.

4.3 SOLID-STATE BATTERIES

Solid-state batteries (SSBs) are described as the next generation of the energy storage industry, which exhibits a set of characteristics highly superior than

Table 4.5 Applications of NiMH Batteries.

Application	*Advantages of NiMH Batteries*
HEVs	Long cycle life, thermal stability, cost-effectiveness
Consumer electronics	Moderate energy density, environmental compatibility
Backup power systems	Overcharge tolerance, reliability

those of conventional Li-ion liquid electrolyte batteries. Indeed, SSBs exploit solid electrolytes to significantly address challenges relating to energy density, longevity, and overall safety, offering a compelling solution for widespread adoption of EVs. This section explains SSB principles and material formulations, performance criteria, challenges, and potential in relation to redefining EVs in the near future.

In contrast to traditional liquid electrolyte-based Li-ion batteries, SSBs are developed with solid electrolytes. The solid electrolytes are ceramics, POEs, or composite materials that exhibit high ionic conductivity; therefore, no leakage, flammability, and thermal runaway typical of liquid electrolytes can occur. The exclusion of a liquid component in SSBs also enhances safety aspects and enables the use of high-performance electrode materials for even better overall battery performance.

One of the most significant advantages of SSBs is that they can achieve a very high energy density, above 400 Wh/kg compared to 150–200 Wh/kg for conventional Li-ion batteries. This has primarily been achieved by allowing SSBs to accommodate Li metal anodes. Li metal has a theoretical capacity of 3,860 mAh/g, which is about ten times that of graphite, the standard anode material in Li-ion batteries. The solid electrolyte is employed as a physical barrier against dendrite formation, a significant concern with Li metal batteries, potentially causing short circuits and thermal events.

Table 4.6 presents the comparison of key properties of SSBs versus liquid electrolyte-based Li-ion batteries. The performance of SSBs demonstrates an obvious advantage of solid-state technology.

As can be seen from Table 4.6, SSBs provide higher performance than conventional Li-ion batteries in terms of energy density, operating temperature range, and cycle life. The high safety margin of SSBs based on their nonflammable solid electrolytes is particularly advantageous for EVs, where thermal management is critical.

The choice of solid electrolyte material greatly influences the performance and applicability of the SSB. One major group, which is gaining more interest today, consists of ceramic-based electrolytes that show strong ionic conductivities along with mechanical strengths, which include lithium lanthanum zirconium oxide (LLZO) and sulfide-based compounds. LLZO's ionic conductivities stand between 0.1 and 1 mS/cm at room temperature

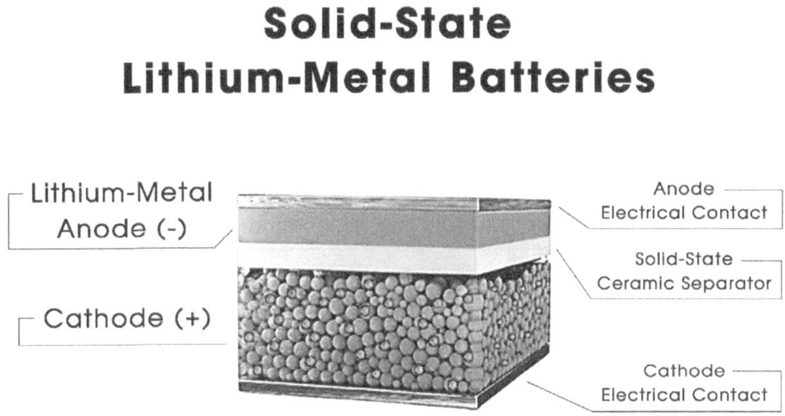

Figure 4.2 SSB Configurations.

Table 4.6 Comparison of SSBs and Li-Ion Batteries.

Parameter	*SSBs*	*Li-Ion Batteries*
Energy density (Wh/kg)	400–500	150–200
Operating temperature range (°C)	−30 to 100	−20 to 60
Cycle life (cycles)	2,000–10,000	1,000–3,000
Safety	High	Moderate

and is equivalent to that of liquid electrolyte. However, its brittle nature makes it challenging to scale up and incorporate with flexible battery designs. Sulfide-based electrolytes possess good ionic conductivity with ductility but are moisture sensitive and require severe conditions for the manufacturing process.

Other solid electrolyte materials used in SSBs include POE electrolytes. Examples are PEOs doped with Li salts. These are flexible, processable but have generally lower ionic conductivities compared with ceramics. Composite electrolytes attempt to achieve optimal levels of the attributes offered by ceramics and POEs for good ionic conductivity and mechanical stability together with appropriate processing. Typical properties of various solid electrolyte materials are presented in Table 4.7.

Table 4.7 summarizes the trade-offs for different solid electrolyte materials. The combination of high conductivity and excellent mechanical robustness makes ceramics ideal; POEs have better processability, while attempts have been made to combine these advantages, promising to be an important area for future research and development.

The performance characteristics of SSBs make them particularly suitable for a range of applications in EVs, where energy density, safety, and

Table 4.7 Properties of Solid Electrolyte Materials.

Material Type	*Ionic Conductivity (mS/cm)*	*Mechanical Strength*	*Processability*
Ceramic (LLZO)	0.1–1.0	High	Low
Sulfide-based	0.1–1.0	Moderate	Moderate
PEO	0.001–0.01	Low	High
Composite	0.01–0.1	Moderate	Moderate

durability are critical. For instance, the long cycle life of SSBs is usually more than 10,000 cycles, ensuring that batteries in EVs can support daily use without significant degradation. Additionally, the wide operating temperature range of SSBs enhances their performance under extreme climates, addressing the limitation of current Li-ion batteries.

Nevertheless, numerous barriers exist toward the possibility of widespread uptake. The high price of materials related to solid electrolytes and complexity of their fabrication process represents an essential barrier that must be overcome. With respect to the synthesis [13–15] of ceramic electrolytes, their sintering process requires a huge amount of energy at the high temperatures, thus leading to relatively high cost. Good contact between the solid electrolyte and electrode materials can also be an issue because gaps and voids can obstruct the ionic transport and reduce efficiency within batteries.

Scalability is another challenge. Although laboratory-scale prototypes of SSBs have shown remarkable performance metrics, scaling up these results to commercial-scale production is a significant hurdle. Issues such as electrolyte uniformity, electrode integration, and manufacturing yield need to be addressed to make mass production cost-effective. Table 4.8 summarizes the key challenges and potential solutions for SSBs.

Table 4.8 presents material, manufacturing, and scaling issues that need to be overcome for full exploitation of SSBs. Material science, engineering, and manufacturing technologies would play a very significant role in overcoming the challenges identified.

Conclusion: SSBs are an advanced generation of energy storage systems. Unmatched energy density, superior safety, and lifespan place SSBs far ahead of other alternatives. Their potential for revolutionizing the EV industry is in addressing traditional Li-ion battery limitations with potential for using next-generation electrode materials. There are various challenges related to cost, manufacturing, and scalability issues that have to be addressed to make SSBs commercially viable. Widespread adoption of SSBs and their transformative impact on sustainable transport will depend on continued research and development alongside strategic investments in manufacturing infrastructure.

Table 4.8 Challenges and Solutions in SSB Development.

Challenge	*Description*	*Potential Solution*
High material costs	Expensive raw materials for solid electrolytes	Development of cost-effective alternatives
Manufacturing complexity	Difficulties in processing and integration	Advanced manufacturing techniques
Interface stability	Formation of interfacial resistance	Coatings and engineered interfaces
Scalability	Challenges in large-scale production	Process optimization

4.4 BATTERY MANAGEMENT SYSTEMS

The most crucial part of the modern EV and other energy storage applications is battery management systems (BMSs), which guarantee safe, efficient, and reliable operation of the battery packs. In battery systems, rising complexity and energy density have made BMSs an essential component to ensure the maintenance of the health of the batteries, optimization of performance, and an extension in the life cycle of batteries. This chapter addresses the architecture, functionality, challenges, and emerging trends of BMSs, discussing their significant role in the whole electric mobility area.

A BMS essentially watches and controls other parameters associated with battery pack- voltage, current, temperature, state of charge (SOC), state of health (SOH), and state of power (SOP). It constantly analyzes all the parameters that may lead it to operate the battery outside the safe operating limits of the battery and mitigate risks of overcharging, over-discharging, thermal runaway, and short circuits. It possesses hardware parts such as sensors, microcontrollers, and communication interfaces along with sophisticated algorithms in software that provide real-time data analysis and decision-making capabilities.

The important function of a BMS is to ensure uniform cell performance of the battery pack. A modern EV battery pack consists of hundreds to thousands of individual cells connected in series and parallel configurations. Because of such interconnections, there will always be an imbalance in cell characteristics, including capacity and internal resistance while charging and discharging. Cell balancing methods are used to eliminate such imbalances. There are two types of balancing techniques: passive and active balancing. The excess energy in passive balancing is dissipated as heat through the resistors. In contrast, active balancing redistributes the energy between cells to balance their state of charge. Table 4.9 presents the comparison of passive and active balancing techniques for cells.

As can be observed in Table 4.9, the energy efficiency with active balancing is higher along with an increase in lifetime of the battery pack but

Table 4.9 Comparison of Passive and Active Cell Balancing Methods.

Parameter	*Passive Balancing*	*Active Balancing*
Energy efficiency	Low (energy dissipated as heat)	High (energy redistributed)
Complexity	Low	High
Cost	Low	High
Cell lifetime impact	Moderate	High (improved longevity)

increases system complexity and also cost. Passive balancing is simple with low cost and could be applied on less stringent applications.

BMSs also offer the function for thermal management. The temperature indeed is a key determining factor of both performance and safety of the batteries. Outside the ideal working temperature range of −15 to 35°C for Li-ion batteries, this results in rapid degradation, inefficiency, and even major safety risks. Temperature is monitored by the BMS with multiple sensors distributed throughout the battery pack. When temperature anomalies are detected, the BMS can trigger activation of cooling or heating systems to keep the operating range intact. Liquid cooling, air cooling, and phase-change materials represent some advances in thermal management. Table 4.10 illustrates the influence of temperature on battery performance metrics.

Table 4.10 shows how temperature will affect battery performance metrics, which explains how to maintain an appropriate temperature for battery safety and life expectancy. The BMS is critical in executing the thermal management strategy, as illustrated below.

State estimation is another functionality of BMSs. Proper calculation of SOC, SOH, and SOP is crucial for effective energy management and for instilling user confidence. SOC estimation determines the remaining charge within the battery and determines charging and discharging. There are mainly two types of SOC estimation techniques, such as Coulomb counting and model-based methods, that make use of complicated algorithms for simulating battery dynamics. SOH estimation measures the health of the battery and predicts its useful life. The SOP of an estimated level determines how much power may be delivered or absorbed by a battery safely. Besides SOP estimation, communication and integration are other activities of BMSs. A BMS provides communication with onboard control systems, charging system of EVs, and any other networking present. Such forms of communication protocols might also appear in the form of a controller area network (CAN) or local interconnect network (LIN). These protocols allow for real-time sharing of critical information, such as battery status and diagnostic data, that enhance the overall functionality and safety of the EV. Table 4.11 presents common communication protocols used in BMSs and their features.

Table 4.10 Impact of Temperature on Battery Performance.

Temperature Range (°C)	*Performance Characteristics*
Below 0°C	Reduced capacity and power output
0–35°C	Optimal performance and longevity
Above 35°C	Accelerated degradation, risk of thermal runaway

Table 4.11 Communication Protocols in BMSs.

Protocol	*Data Rate (kbps)*	*Key Features*	*Applications*
CAN	Up to 1,000	High reliability, error detection	EV battery management
LIN	Up to 20	Low cost, simple implementation	Low-power subsystems
Universal asynchronous receiver-transmitter (UART)	Variable	Versatile, easy integration	Data logging, diagnostics

Despite their critical importance, BMSs face several challenges. The key challenge would be the trade-off between cost and performance. Implementing high-performance BMS architectures with active balancing, for example, and sophisticated state estimation algorithms is very costly. Process cost might be a challenge for achieving mass-market EVs. The second is that integrating BMSs with newer battery chemistries like SSBs and Li–sulfur batteries may also pose some challenges. These technologies exhibit different electrochemical behaviors and, thus, necessitate new algorithms and hardware designs to be used.

The improvements in BMS technology are focused on such problems. First, machine learning and artificial intelligence (AI) have been integrated into the BMS algorithm to increase the accuracy in state estimation and predictive maintenance. AI-driven models can process vast amounts of data and recognize patterns or anomalies, making real-time diagnostics and proactive interventions possible. Second, wireless BMSs have recently gained popularity, owing to the minimal wiring complexity and weight within a battery pack. The BMSs monitor cells wirelessly, which has streamlined the design and packaging of large battery packs.

Summary: Modern battery packs cannot exist without a BMS for assurance about safety and efficiency. Their functions range from basic monitoring to cell balancing, thermal management, state estimation, and communication, all of which enhance the performance and lifespan of a battery. However, cost, integration, and scalability issues still linger. But materials, algorithms, and communication technologies keep driving BMSs. This would further open up BMS technology as central to the achievement of efficient widespread adoption of EVs and other battery-powered systems based on their growing demand for reliable high-performance energy storage solutions.

4.5 CHALLENGES AND FUTURE TRENDS IN BATTERY TECHNOLOGY

The battery is, in fact, the heart of EVs, portable electronics, and renewable energy systems. After such great progress, however, many challenges remain to really unlock the full potential of the sector. Among the biggest ones are material availability, energy density limits, cost barriers, inefficiencies in recycling, and concerns about safety. The future of energy storage is promised through the rising trends of SSBs, next-generation materials, advanced manufacturing techniques, and smart battery systems. These challenges and future trends of impact in battery technology and electric mobility are what the author writes about.

The most direct challenge for battery technology lies in the availability of raw critical materials, such as Li, cobalt, and nickel. The global demand for Li is expected to grow from 82,000 metric tons in 2021 to over 700,000 metric tons by 2030, led by increasing adoption of EVs. Cobalt supply risks also remain as high-energy cathodes because of the limited geographical distribution of cobalt: the Democratic Republic of Congo holds more than 60% of the global production. These vulnerabilities in the supply chain include a source of price volatility and resource unsustainability. Table 4.12 provides ten years of projected demand for primary battery materials.

The data provided in Table 4.12 above shows the exponential growth of the demand for Li, cobalt, and nickel but highlights the need for more sustainable alternatives and material recycling.

Energy density is another issue related to existing battery technologies. While the Li-ion batteries have succeeded in achieving energy densities of 150–200 Wh/kg, more will be needed to achieve the performance required for long-range EVs and grid-scale energy storage. Another promising area is SSBs, where liquid electrolytes are replaced by solid alternatives. This could achieve an energy density above 400 Wh/kg because of compatibility with

Table 4.12 Projected Demand for Key Battery Materials (in Metric Tons).

Material	*2021*	*2025*	*2030*
Lithium	82,000	300,000	700,000
Cobalt	140,000	200,000	400,000
Nickel	1,400,000	1,800,000	2,500,000

Li metal anodes. However, the complexity of manufacturing and high costs are the significant barriers to commercialization.

The cost of the battery is another barrier to the widespread adoption of EVs. Even though the cost of Li-ion batteries has come down from $1,200 per kWh in 2010 to around $150/kWh in 2022, the break-even point for EVs to reach cost parity with internal combustion engine vehicles in the process is at $100/kWh. Improvements in materials, manufacturing processes, and economies of scale can yield cost reductions. Cobalt-free cathodes with LFP-style compositions have already begun material cost decreases while also improving on safety and cycle life.

The principal challenge with high-energy battery systems relates to their safety. Thermal runaway due to the nature of an internal short circuit and overheating tends to cause fires and even explosions. Due to such incidents related to the incidence of fire in EVs and other portable products, safety issues have taken over the limelight. The techniques developed include advanced thermal management systems, nonflammable electrolytes, and circuits for overcharge protection. Table 4.13 presents the common safety issues and mitigation strategies in battery technology.

Table 4.13 illustrates that addressing safety concerns requires a combination of advanced materials, robust engineering, and intelligent control systems.

Recycling Inefficiency: The Li-ion battery recycling rate is below 10% globally. Because of this inefficiency, resources and the environment are also wasted. Traditional methods of recycling occur through pyrometallurgy and hydrometallurgy. These have enormous energy requirements and very poor recoveries. Emerging recycling technologies, such as direct recycling, recover materials from used batteries in a usable form with energy savings and without negative environmental impacts. An important milestone in the development of a circular economy for batteries is a cost-effective and scalable recycling process.

Several trends will affect battery technology over the coming years. The most dramatic and consequential transition, however, will be the transition

Table 4.13 Safety Concerns and Mitigation Strategies in Battery Technology.

Safety Concern	*Cause*	*Mitigation Strategy*
Thermal runaway	Internal short circuit, overheating	Advanced thermal management, nonflammable electrolytes
Overcharging	Excessive voltage	Overcharge protection circuits
Dendrite formation	Li plating	Solid electrolytes, optimized charging protocols
Material degradation	High temperatures, cycling	Enhanced material design

to SSBs. SSBs should offer much higher energy density and much higher safety over a traditional Li-ion battery, and potentially longer cycle life as well. Companies like Toyota, Solid Power, and QuantumScape are accelerating the commercialization of SSBs, and prototypes will reach the market in just a few years.

The focus is now on next-generation materials, especially on Li–sulfur and Li–air chemistries. Li–sulfur can provide a theoretical energy density of 2,600 Wh/kg, which is considerably more than that of Li-ion batteries. However, the polysulfide shuttle effect and limited cycle life remain challenges to fully realize the former. In a similar manner, it can be said that with respect to Li–air with an energy density of about 3,500 Wh/kg, oxygen stability and electrode degradation are the main issues.

High-performance manufacturing techniques, such as roll-to-roll processing, and 3D printing are making battery production scalable and cost-effective. Such methods have allowed for electrodes with reduced thickness, meaning improved energy density and charge rates. AI and machine learning will be integrated into the design and management systems of the battery.

Another significant trend is smart battery systems with sensors and advanced control algorithms that monitor and optimize in real time and allow predictive maintenance and enhancing safety. For instance, Tesla's BMS relies on machine learning to calculate degradation and adapt charging protocols to maximize the life of a battery.

Table 4.14 presents the critical challenges and future trends in battery technology.

Interlacing challenges and innovations best, Table 4.14 underpins the necessity of holistic approaches to pushing out the battery technology.

Table 4.14 Key Challenges and Future Trends in Battery Technology.

Challenge/Trend	*Description*	*Implication*
Material scarcity	Limited availability of Li, cobalt, and nickel	Focus on alternative materials and recycling
Energy density limitations	Constraints on range and storage capacity	Development of solid-state and next-gen chemistries
Cost barriers	High production costs	Economies of scale, advanced manufacturing
Safety concerns	Risk of thermal runaway and fires	Improved thermal management, robust BMS
Recycling inefficiencies	Low recovery rates	Development of efficient recycling technologies
SSBs	High energy density, improved safety	Revolutionize EV and grid storage
Next-generation materials	Li–sulfur, Li–air chemistries	Breakthroughs in energy density

The landscape characterizing a challenge and an opportunity when it comes to battery technologies entails various challenges and opportunities. As an instance, the scarcity in resources, energy density, or problems with cost, safety concerns, and inefficiency regarding recycling will be resolved at a coordinated level between doing research, industry, or policy. The trends in SSBs, next-generation materials, advanced manufacturing, and smart systems also highlight an impressive energy storage future that will be safer, more efficient, and sustainable. The better the battery technology continues to improve, the more it will play an increasingly important role in allowing a transition toward a low-carbon economy and sustainable energy systems.

REFERENCES

1. Kek, V., Kandasamy, J., Vezhavendhan, R. *et al.* Digital technology adoption model for electric vehicle battery recycling supply chain—an influential relationship mapping. *Oper Manag Res* 17, 1469–1508 (2024). https://doi.org/10.1007/s12063-024-00515-y
2. Cheng, A.L., Fuchs, E.R.H., & Michalek, J.J. US industrial policy may reduce electric vehicle battery supply chain vulnerabilities and influence technology choice. *Nat Energy* 9, 1561–1570 (2024). https://doi.org/10.1038/s41560-024-01649-w
3. Abdullah, M., Moin, L., Ahmed, F. *et al.* Electric vehicle battery temperature control using fuzzy logic. *Aut Control Comp Sci* 58, 237–251 (2024). https://doi.org/10.3103/S0146411624700135
4. Satrio, J., Juned, M., & Salam, S. International and domestic factors of battery electric vehicle technology diffusion in Japan. *East Asia* 41, 109–127 (2024). https://doi.org/10.1007/s12140-023-09418-4
5. Parthasarathy, T.N., Narayanamoorthy, S., Devi, N.S.K. *et al.* An idiosyncratic interval valued picture q-rung orthopair fuzzy decision-making model for electric vehicle battery charging technology selection. *Int J Fuzzy Syst* 26, 2023–2038 (2024). https://doi.org/10.1007/s40815-024-01683-6
6. Gao, C., Zhao, X., Zhu, W. *et al.* Impact load between spherical obstacles and electric vehicle battery packs. *J Mech Sci Technol* 38, 2845–2854 (2024). https://doi.org/10.1007/s12206-024-0504-3
7. Zhu, T., Boyles, S.D., & Unnikrishnan, A. Battery electric vehicle traveling salesman problem with drone. *Netw Spat Econ* 24, 49–97 (2024). https://doi.org/10.1007/s11067-023-09607-x
8. Liu, Q., Zhang, Z., & Zhang, J. Research on the interaction between energy consumption and power battery life during electric vehicle acceleration. *Sci Rep* 14, 157 (2024). https://doi.org/10.1038/s41598-023-50419-3
9. Dhairiyasamy, R., Gabiriel, D., Bunpheng, W. *et al.* A comprehensive analysis of India's electric vehicle battery supply chain: Barriers and solutions. *Discov Sustain* 5, 361 (2024). https://doi.org/10.1007/s43621-024-00595-7
10. Elbaz, H., Bourzik, M., & Elhilali Alaoui, A. Planning a sustainable electric vehicle infrastructure considering battery life: Modeling and resolution by the multi-agent Q-learning metaheuristic. *Arab J Sci Eng* 49, 12537–12559 (2024). https://doi.org/10.1007/s13369-024-08741-z

11. Pinto-Bautista, S., Baumann, M., & Weil, M. Prospective life cycle assessment of an electric vehicle equipped with a model magnesium battery. *Energ Sustain Soc* 14, 44 (2024). https://doi.org/10.1186/s13705-024-00475-y
12. Geetha, A., Suprakash, S., & Lim, SJ. Sensor based battery management system in electric vehicle using IoT with optimized routing. *Mobile Netw Appl* 29, 349–372 (2024). https://doi.org/10.1007/s11036-023-02262-z
13. Lyu, W., Hu, Y., Liu, J. *et al.* Impact of battery electric vehicle usage on air quality in three Chinese first-tier cities. *Sci Rep* 14, 21 (2024). https://doi.org/10.1038/s41598-023-50745-6
14. Kim, G., Kang, S., Park, G. *et al.* Electric vehicle battery state of charge prediction based on graph convolutional network. *Int J Automot Technol* 24, 1519–1530 (2023). https://doi.org/10.1007/s12239-023-0122-6
15. Vijaya Saraswathi, R.J., Krishnakumar, V., Vasan Prabhu, V. *et al.* Hybrid energy management strategy for ultra-capacitor/battery electric vehicles considering battery degradation. *Electr Eng* (2024). https://doi.org/10.1007/s00202-024-02533-2

Chapter 5

Charging Infrastructure

5.1 HOME CHARGING STATIONS

Home charging stations are crucial for mass adaptation of electric vehicles (EVs) as users will be able to charge up their EVs conveniently and effectively. Therefore, with the number of electric cars on the roads now, they need to be properly charged for a smooth day. Reliability and efficiency have become the new standards in the EV ecosystem; hence, the focus of this chapter is on the technical specifications and cost implications for installation purposes and general implications on adopting electric mobility.

Home charging stations are also known as electric vehicle supply equipment (EVSE). They come in two main types: Level 1 and Level 2 chargers. Level 1 chargers use a standard 120-volt electrical outlet and can be used by most households without additional installations. These chargers typically deliver power at a rate of 1.4 kW, which translates to approximately 4–5 miles of range per hour of charging. This would be sufficient for overnight charging if the daily commute is modest, typically under 40 miles. For people who travel farther or require quicker charging, Level 2 chargers are ideal.

Level 2 chargers use a 240-volt circuit. Level 2 home chargers [1–3] range from 3.3 kW to 19.2 kW based on the model and the electrical system. Level 2 home chargers can provide 20–60 miles of range for each hour, making it possible to charge significantly more within the same time. Key specifications of Level 1 and Level 2 home chargers are provided in Table 5.1.

Installing a home charging station is very detailed, considering the variety of things that have to be looked at. Homeowners generally have to upgrade their electrical system in order to be able to accommodate the higher power requirements for Level 2 chargers. The installation usually includes the provision of a dedicated 240-volt circuit. Such chargers range in cost from $500 to $2,000 depending on the complexity of the job and the local labor rates. Additionally, the location of the charging station should be convenient and safe, which is often a garage or a driveway. It is also important to ensure that the installation complies with the local electrical codes and standards to avoid any potential hazards.

DOI: 10.1201/9781003608097-5

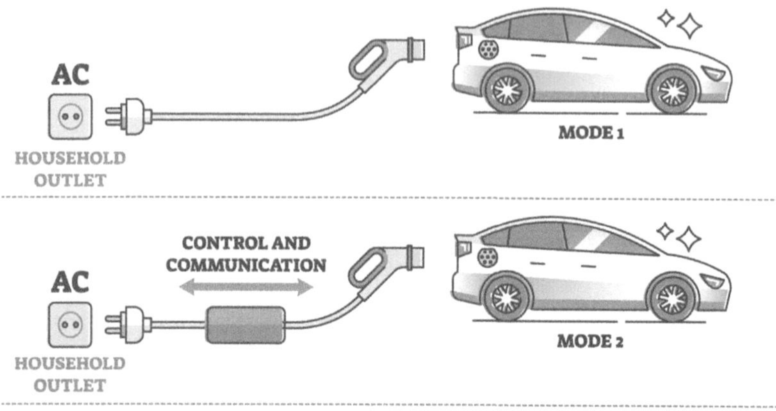

Figure 5.1 EV Charging Modes.

Table 5.1 Comparison of Level 1 and Level 2 Home Charging Stations.

Parameter	*Level 1 Charger*	*Level 2 Charger*
Voltage	120 volts	240 volts
Power output	1.4 kW	3.3–19.2 kW
Charging speed	4–5 miles/hour	20–60 miles/hour
Installation cost	$0–$200 (uses existing outlet)	$500–$2,000 (requires circuit upgrade)
Suitability	Low-mileage users	High-mileage users or those needing faster charging

The cost of the charging equipment varies greatly depending on the features provided. Basic Level 2 chargers with minimal features can be had for around $300, but more advanced models, which include Wi-Fi capability, smartphone apps, and the scheduling of charging based on the electricity rate, can cost up to $1,200. Advanced features allow for optimal scheduling of charge cycles that minimize energy consumption. Table 5.2 illustrates cost breakdowns for a home charging station.

The convenience factor is one of the most important advantages of home charging. Owners can simply park their vehicle in the parking lot overnight and wake up with a fully charged battery to go, thus saving time and decreasing reliance on outside infrastructure, which may not be readily available at all times. Recharging at home tends to be much cheaper compared to using public charging stations. This tends to be much cheaper with a combination of a time-of-use electricity rate. For instance, according

Table 5.2 Cost Breakdown for Home Charging Station Installation.

Component	*Cost Range*
Level 1 charger equipment	\$0–\$200
Level 2 charger equipment	\$300–\$1,200
Electrical system upgrade	\$500–\$2,000
Professional installation	\$200–\$1,000
Total estimated cost	\$700–\$4,200

Table 5.3 Solar-Powered Home Charging Station Analysis.

Parameter	*Value*
Solar panel capacity	5 kW
Daily energy generation	~20 kWh
Average EV consumption	30 kWh/100 miles
Daily driving range	60–70 miles
Annual CO_2 reduction*	~3,600 kg

*Assumes grid electricity mix emits 0.45 kg CO_2 per kWh.

to data from the United States, the average household cost of electricity is about \$0.13/kWh. Assuming an EV consumes about 30 kWh to travel 100 miles, charging at home will cost about \$3.90 for 100 miles, much cheaper than fueling a gasoline car.

The other advantages of home charging stations have to do with energy efficiency and environmental benefits. Homeowners can reduce their carbon footprint even more by using renewable sources like solar panels and reduce operational costs. A home with a 5-kW solar array will generate about 20 kWh of electricity daily when the sun shines bright. This can power an EV for 60–70 miles of driving per day. Table 5.3 outlines the savings and environmental benefits possible using this integrated solar-powered home charging approach.

However, several challenges persist: installing a Level 2 home charging station often presents barriers in the minds of potential EV buyers who reside in leased apartments or multi-unit residences. It is also limiting home charging in areas with very densely populated urban spaces where accessible and safe parking spaces are a major constraint. Governments and utility companies have come up with various kinds of incentives to overcome these tasks, such as tax credits, rebates, and subsidies. For example, in the United States, the federal government provides a tax credit for residential EVSE installations and scenarios that is up to 30% of the installation cost, capped at \$1,000 for different scenarios.

Home charging stations are the heart of the EV ecosystem, which makes recharging practical and efficient for the owners of the EVs and scenarios. It all will depend on users' need to drive, budget, and availability of good electrical infrastructure so that they can choose Level 1 over Level 2 charging. But the story of continued adoption of EVs will only begin when the home charging technology continues to evolve [4–6] and supportive policies start to bridge the current gaps so that the whole experience of EV ownership is further improved.

5.2 PUBLIC CHARGING STATIONS

Public charging stations are also an integral part of the EV ecosystem because they make people feel less anxious about going out of range and provide access to basic charging options to those without a home or a workplace within easy reach. Public charging infrastructure will move EV users inside towns, cities, highways, and rural areas and support longer-distance trips as well, since it's useful for everyone who will use shared parking on-street or off. This section addresses the categorization, deployment strategies, economic effects, and obstacles surrounding public charging stations.

Public charging stations can primarily be grouped into Level 2 chargers and direct current fast chargers, sometimes referred to as DCFCs. Level 2 public chargers draw charge from a 240-volt circuit like that found in a home but deliver a greater power range, 6.6–19.2 kW. They typically deliver 20–30 miles of range per hour of charging and would be most appropriate for users who spend longer periods of time out of home, such as at shopping malls, work locations, or residences. In contrast, DCFCs provide much

Figure 5.2 Public Charging Stations.

higher power levels, typically in the range of 50–350 kW, which allow EVs to be charged rapidly. A typical DCFC can deliver 80% of battery capacity in 20–30 minutes, which is suitable for highway rest stops and urban fast-charging hubs. Table 5.4 compares the main features of Level 2 chargers and DCFCs in public charging scenarios.

There are several strategic considerations when installing public charging stations. Locations are chosen based on traffic patterns, proximity to amenities, and accessibility. Urban centers prioritize fast-charging hubs near commercial districts to cater to commuters and ride-hailing services. In contrast, highway corridors emphasize high-capacity DCFC networks to enable long-distance travel. Governments and private companies often collaborate to identify optimal sites and deploy infrastructure efficiently. For instance, the United States [7–9] launched its National Electric Vehicle Infrastructure (NEVI) program to award funds for filling "gaps" in infrastructure along highways with public-access charging to fill in those gaps.

For public charging stations, the cost is relatively higher than that of home charge points as there are ancillary requirements, such as changing electrical grids, modifying parking spaces for the installation, and fitting safety standards. Level 2 chargers cost between $3,000 and $7,000 to install per unit, while the DCFC cost starts at $25,000 and can go up to $100,000 in some cases, depending upon the power output and site conditions. These costs are often subsidized either by government incentives or between private entities and public agencies. Table 5.5 provides a detailed breakdown of the installation cost for public charging stations.

Table 5.4 Comparison of Level 2 Chargers and DCFCs.

Parameter	*Level 2 Charger*	*DCFC*
Voltage	240 volts	400–800 volts
Power output	6.6–19.2 kW	50–350 kW
Charging speed	20–30 miles/hour	100–250 miles/hour
Typical usage time	2–8 hours	20–30 minutes
Installation cost	$3,000–$7,000	$25,000–$100,000
Best-suited locations	Workplaces, retail spaces	Highways, urban fast-charging hubs

Table 5.5 Cost Components for Public Charging Station Installation.

Component	*Level 2 Charger ($)*	*DCFC ($)*
Charging equipment	500–2,500	10,000–50,000
Electrical infrastructure	1,500–3,000	10,000–25,000
Installation and labor	1,000–1,500	5,000–20,000
Maintenance and upgrades	500/year	2,000/year
Total estimated cost	3,000–7,000	25,000–100,000

Public charging stations are an extremely important factor in reducing range anxiety, which is still one of the biggest barriers to the adoption of EVs. A strong and well-distributed network gives drivers confidence that they will be able to recharge easily, even on long trips. Public chargers are thus installed in nearly all high-density population centers of cities and towns, in parking lots and street curbs, and near retail stores so that drivers can charge while shopping or dining. DCFCs are installed along highways at distances of 50–70 miles so that long distances can be covered without running out of charge.

The economic model for public charging stations is pay-per-use, subscription plans, or a combination of both. The cost of charging would depend on the location, power output, and electricity cost. For instance, Level 2 chargers charge around $0.10–$0.30/kWh, while DCFCs charge $0.20–$0.60/kWh. Most networks also impose session fees or idle fees to discourage users from occupying charging spaces once fully charged. Table 5.6 presents the comparison of the cost of charging an EV at public and residential stations.

All these developments have attracted investments from private companies and startups. Operators—Tesla, ChargePoint, Electrify America, etc.—are also expanding their networks [10–12] all around the globe, providing services like mobile applications, reservation systems, etc., compatible with different EV models. Altogether this enhances user experience, thereby making the adoption of EVs easy and hassle-free.

Despite these advances, many barriers to public charging infrastructure deployment exist. High upfront costs and long permitting processes, along with issues of grid capacity, are some of the main challenges. Lastly, equal access to chargers in rural and underserved areas is essential for widespread adoption. Governments worldwide have addressed these issues through funding programs, streamlined regulations, and partnerships with utility companies.

In conclusion, public charging stations form an integral part of the EV ecosystem as a support for a sustainable transportation system. They fill any gap that may exist between the home and the road of EV users who do not have access to home charging. This consequently means they can have long-distance travel capacity of the car. Public charging infrastructure is rapidly

Table 5.6 Cost of Charging an EV at Public and Residential Stations.

Parameter	*Residential Charging ($/kWh)*	*Public Level 2 Charging ($/kWh)*	*Public DC Fast Charging ($/kWh)*
Average cost	0.13	0.20	0.40
Cost for 100 miles (30 kWh)	3.90	6.00	12.00
Charging time (hours)	8–10	2–6	0.5–1

evolving to enable electric mobility with strategic investments, innovations in technologies, and planning. Cost, accessibility, and grid integration challenges must be approached collectively to ensure the expansion of these networks is a success.

5.3 FAST-CHARGING NETWORKS

Fast-charging networks are at the heart of any modern EV infrastructure, as they support the important need for rapid charging to be able to travel far and avoid downtime. Networks are usually based on DCFCs that supply large amounts of power directly to the vehicle battery, drastically reducing charging time compared with Level 1 and Level 2 chargers. This section takes a deep dive into the fast-charging network technology, the deployment strategy, the challenges it faces, and its role in supporting EV adoptions.

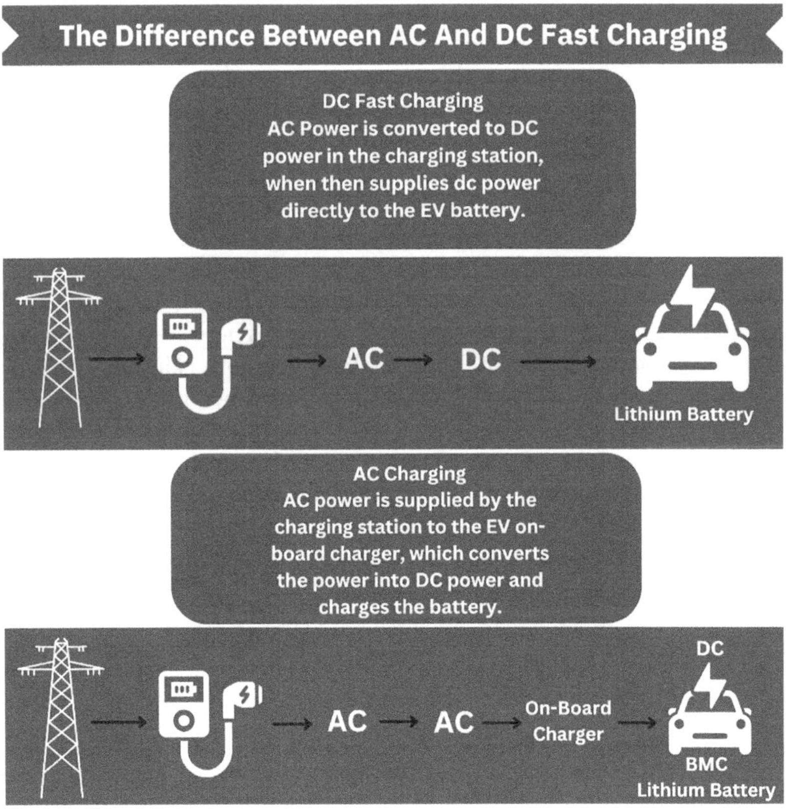

Figure 5.3 Fast-Charging Networks.

Fast-charging networks are characterized by the ability to provide power between 50 kW and over 350 kW, depending on the charger specifications and the abilities of the EV battery. These chargers work on a high voltage system, usually between 400 and 800 V, but advanced networks even look at 1,000 V as a potential target to gain charging speed. Such chargers can add 150 miles in less than 30 minutes using 50 kW DC fast charge or less than 10 minutes using 350 kW ultra-fast charge. Fast charging networks are essential for travel between cities and in urban regions with heavy traffic, due to their effectiveness and ease of use.

The implementation of fast-charging networks involves heavy technical and logistical considerations. Chargers need high-tech cooling systems to cope with the heat generated during the transfer of high power, keeping safety and efficiency in mind. In addition, fast-charging networks require strong connections to the grid that could handle peak power demands. In addition, renewable energy sources like solar or wind power are increasingly becoming part of fast-charging stations to improve their sustainability and reduce their dependence on the grid. Table 5.7 summarizes key specifications and performance metrics of typical fast chargers.

The main motivation for deploying a fast-charging network is the provision of reliable access to widespread rapid charging. The high-priority focus includes highways with chargers at an interval of 50–70 miles, allowing easy long-distance travel. Those locations that are likely to have high penetration of EVs and locations that will not offer much opportunity for home charging will be focused. Governments, in most cases, partner with private firms to raise and apply the funds. For instance, the European Union and the United States are implementing the TEN-T initiative and the NEVI program that would install fast-charging corridors throughout every state.

Installing a fast-charging station is expensive. The usual DCFC installation can cost between $25,000 and $50,000 per unit. The ultra-fast charger stations cost as low as $100,000–$150,000 per unit. These include site preparation, electrical upgrades, grid connections, and the hardware itself. Among the cost components are operational ones such as maintenance and acquiring energy. However, through charging fees and incentives,

Table 5.7 Key Specifications of Fast Chargers.

Parameter	*Standard DCFC*	*Ultra-Fast Charger*
Power output	50–150 kW	150–350+ kW
Voltage range	400–500 volts	800–1,000 volts
Charging speed	150 miles/30 minutes	150 miles/10 minutes
Typical locations	Highways, urban hubs	Major highways, premium locations

governments can offset operational costs. Table 5.8 breaks down the cost components associated with the installation of fast-charging stations.

Charging fees, station availability, and vehicle compatibility are determinants of the usage of fast-charging networks. Generally, charging costs range from \$0.20 to \$0.60 per kWh. Some networks introduce additional session fees or idle fees to optimize usage at stations. Standard connectors include the combined charging system (CCS), CHAdeMO, and Tesla's proprietary Supercharger connector, which will allow compatibility with many different models of EVs. One of the biggest challenges remains the absence of a universal standard in regions where markets for vehicles are diverse.

Integration of renewable energy and energy storage systems is essential in fast-charging networks for reducing the carbon footprint of EVs. A solar panel and a wind turbine, integrated into charging stations, supply partial energy, and the stored systems in the batteries provide grid stability to avoid peak loads. As an example, a 100-kW solar panel array produces around 400 kWh per day and can charge 10–12 EVs. Table 5.9 presents the potential environmental benefits of integrating renewable energy sources with fast-charging stations.

Despite these benefits, challenges exist for fast-charging networks. High installation and maintenance costs are a challenge, especially in less-populated or underserved areas. The need for high-capacity grid connections strains local infrastructure, which may have to be upgraded and contribute

Table 5.8 Cost Breakdown for Fast-Charging Station Installation.

Component	*Standard DCFC (\$)*	*Ultra-Fast Charger (\$)*
Charging equipment	10,000–20,000	50,000–100,000
Site preparation	5,000–10,000	10,000–20,000
Electrical upgrades	5,000–15,000	20,000–40,000
Grid connection	5,000–10,000	10,000–30,000
Maintenance (annual)	1,000–3,000	5,000–10,000
Total estimated cost	25,000–50,000	100,000–150,000

Table 5.9 Environmental Benefits of Renewable Energy Integration.

Parameter	*Value*
Solar array capacity	100 kW
Daily energy generation	~400 kWh
EVs charged per day	10–12
Annual CO_2 reduction*	~65,700 kg

* Assumes grid electricity mix emits 0.45 kg CO_2 per kWh.

to the overall cost. Ensuring that equitable access to fast chargers exists for all users of EVs, including those in low-income communities, is also an ongoing challenge.

Government policies and incentives will help overcome these barriers. Some of the common incentives include subsidies for the installation of stations or tax credits for operators, and money in the form of grants for grid upgrades. For instance, the US government offers up to 30% tax credits for the installation of EV charging stations, while Norway and China offer massive subsidies to speed up network establishment.

In summary, fast-charging networks will be a critical component to supporting the transition to electric mobility since they meet the urgent demand for rapid and reliable charging solutions. Their potential to recharge EVs promptly and efficiently makes them highly essential for long-distance and urban mobility. Although cost, grid integration, and accessibility [13–15] remain among the challenges, advancements in technology, as well as favorable policies, are providing access to growth for fast-charging infrastructure. The convenience and practicality of owning an EV are bound to be enhanced by further promoting fast-charging networks by facilitating collaboration among governments, private companies, and energy providers.

5.4 CHARGING STANDARDS AND COMPATIBILITY

Charging standards and compatibility are integral parts of EV infrastructure, making sure that charging stations interoperate with EVs of any manufacturer. The development of universal and region-specific standards facilitates seamless experiences for users and supports broader adoption of EVs. This section examines the main charging standards, connector types, protocols, and challenges associated with maintaining compatibility in a fast-evolving industry.

Charging standards can be categorized into two categories: by power level and type of current. Alternating current (AC) charging standards are the most commonly used with Level 1 and Level 2 chargers at home and in public places, while DC charging standards are usually used for fast charging. The AC charging standards include Type 1 (SAE J1772) and Type 2 (IEC 62196). Type 1 is the most common in North America and Japan. Type 2 is used in most of Europe and other countries. DC charging standards include CCS, CHAdeMO, and Tesla's proprietary Supercharger standard. Technical specifications and geographical usage for these charging standards are presented in Table 5.10.

Compatibility is the most critical parameter in charging infrastructure design and deployment. The compatible connector types and communication protocols need to be accepted by vehicles and chargers for charging purposes. For example, if CCS is the dominant standard in a given region, auto

Table 5.10 Overview of EV Charging Standards.

Standard	*Connector Type*	*Current Type*	*Power Output*	*Geographical Usage*
Type 1 (SAE J1772)	Single-phase AC	AC	Up to 7.2 kW	North America, Japan
Type 2 (IEC 62196)	Single- and three-phase AC	AC	Up to 43 kW	Europe, Australia, Asia
CCS	Combined Type 1/2	DC	50–350 kW	Global
CHAdeMO	Unique	DC	Up to 400 kW	Japan, Europe
Tesla Supercharger	Proprietary	DC	Up to 250 kW	Tesla-specific (global)

manufacturers make sure that the CCS ports of their cars are compatible to enable public charging facilities. On the contrary, CHAdeMO connectors are the norm on Japanese EVs such as those from Nissan and Mitsubishi. In the United States, Tesla utilizes a proprietary connector, whereas it uses CCS adapters in Europe in order to satisfy the regulations found in that country.

Communication protocols are required in order to function as a form of controllers in the flow of power and ensure safety. The first such standard protocol is open charge point protocol (OCPP), which ensures that compatibility is achieved between charging points and management systems, allowing for remote monitoring and load balancing. Another system, the ISO 15118, is called the Plug & Charge system. The system makes it easy for vehicles to authenticate automatically and initiate charging without further user input or payment card. This simplifies the whole user experience and boosts compatibility with EVs for charging networks. Table 5.11 summarizes key communication protocols used in charging EVs.

Charging standard and its compatibility have hugely impacted worldwide EV market trends. Universal standards like CCS facilitate easier and cost-effective car manufacturing by providing a common interface. Regional standards, like the Type 2 used in Europe, provide for conformity to particular area-specific grid systems and applicable regulations. Even so, different standards often present users and infrastructure providers with problems when one standard does not dominate in a specific geographical area.

The biggest challenge facing universality is the expense of implementing multiple connector types and communication protocols into one charging station. For example, a high-powered charging station that accepts CCS, CHAdeMO, and Tesla's Supercharger standards would add more hardware and possibly software, making installation and service work more expensive. Complexity would cause user confusion and inefficiencies. Drivers may come to find stations that are not compatible with their vehicle.

Table 5.11 Key Communication Protocols in EV Charging.

Protocol	*Functionality*	*Application*
OCPP	Charger management, remote monitoring	Public and private charging networks
ISO 15118	Plug & Charge, bidirectional communication	EV-to-grid and smart charging
CHAdeMO	Communication between vehicle and charger	DC fast charging
CCS	Unified AC/DC communication	Global DC fast charging standard

Another challenge that comes from the rapid pace of innovation in battery and charging technology is the requirement of the charging standard. Increasing power levels in EV batteries necessitates upgrading the charging standard for even faster charging. In this way, ultra-fast charging is upgraded to 800 volts as opposed to the 400 volts for fast chargers, where all connectors and communication protocols will be upgraded. Table 5.12 gives an example of charging performance and infrastructure requirements on various voltage levels.

To overcome these challenges, governments, industry bodies, and manufacturers are working toward bringing greater standardization and collaboration for a better future. Initiatives like the CCS-promoting CharIN consortium have come up to establish a global standard in the case of EV charging. To make it more user-friendly at the public charging stations, even the European Union has mandated Type 2 and CCS connectors. In the United States, Tesla has announced its plan to open up its Supercharger network to non-Tesla EVs, signaling a shift toward greater compatibility.

The future of charging standards and compatibility lies in smart charging and vehicle-to-grid (V2G) technologies. Such innovations will facilitate bidirectional energy transfer, meaning that EVs will become mobile energy storage units, and can support grid stability during peak demand. Standards such as ISO 15118 will be important to enable such functionalities, opening up a more sustainable and efficient energy ecosystems.

In brief, standards of charging and compatibility act as a foundation for the growth and success of the EV industry. Universally and region-specific developed standards have been helping popularize the wide usage of EVs. Improvements in communication protocols enhance user experience and improve efficiencies in the system. Despite these challenges within technological evolution and co-existence of multiple standards, there is still much effort within harmonization and innovation to further make the industry compatible and sustainable. Ongoing collaboration from all stakeholders will help address the challenges and, in turn, ensure that charging infrastructure continues to adapt and improve as the rapidly growing EV market demands.

Table 5.12 Impact of Voltage Levels on Charging Performance.

Voltage Level	*Power Output Range*	*Charging Speed (150 miles)*	*Infrastructure Requirements*
400 volts	50–150 kW	~30 minutes	Standard grid connection
800 volts	150–350+ kW	~10 minutes	Enhanced grid capacity, cooling
1,000 volts	>350 kW	<10 minutes	Advanced infrastructure upgrades

REFERENCES

1. Kou, G., Eti, S., Yüksel, S. *et al.* Innovative solution suggestions for financing electric vehicle charging infrastructure investments with a novel artificial intelligence-based fuzzy decision-making modelling. *Artif Intell Rev* 58, 16 (2025). https://doi.org/10.1007/s10462-024-11012-w
2. Horesh, N., Trinko, D.A., & Quinn, J.C. Comparing costs and climate impacts of various electric vehicle charging systems across the United States. *Nat Commun* 15, 4680 (2024). https://doi.org/10.1038/s41467-024-49157-5
3. Sayarshad, H.R. Optimization of electric charging infrastructure: Integrated model for routing and charging coordination with power-aware operations. *NPJ Sustain Mobil Transp* 1, 4 (2024). https://doi.org/10.1038/s44333-024-00004-6
4. Baek, K., Lee, E., & Kim, J. A dataset for multi-faceted analysis of electric vehicle charging transactions. *Sci Data* 11, 262 (2024). https://doi.org/10.1038/s41597-024-02942-9
5. Ji, J., Liu, Y., & Yang, C. Dual ant colony optimization for electric vehicle charging infrastructure planning. *Appl Intell* 53, 26690–26707 (2023). https://doi.org/10.1007/s10489-023-04772-5
6. Mohammed, A., Saif, O., Abo-Adma. M.A. *et al.* Multiobjective optimization for sizing and placing electric vehicle charging stations considering comprehensive uncertainties. *Energy Inform* 7, 131 (2024). https://doi.org/10.1186/s42162-024-00428-x
7. Bhatti, A.R., Tamoor, M., Liaqat, R. *et al.* Electric vehicle charging stations and the employed energy management schemes: A classification based comparative survey. *Discov Appl Sci* 6, 503 (2024). https://doi.org/10.1007/s42452-024-06190-9
8. Kim, S., Yoo, Y., Choi, M. *et al.* The effect of dynamic wireless charging systems on electric vehicle activation and sustainability. *Clean Techn Environ Policy* (2024). https://doi.org/10.1007/s10098-024-03043-2
9. Elbaz, H., Bourzik, M., & Elhilali Alaoui, A. Planning a sustainable electric vehicle infrastructure considering battery life: Modeling and resolution by the multi-agent Q-learning metaheuristic. *Arab J Sci Eng* 49, 12537–12559 (2024). https://doi.org/10.1007/s13369-024-08741-z
10. Alanazi, A., Jan, S.T., Alanazi, M. *et al.* Analysis of hybrid energy systems for electric vehicle charging of different demographics. *Clean Techn Environ Policy* (2024). https://doi.org/10.1007/s10098-024-02878-z

11. Lou, J., Shen, X., Niemeier, D.A. *et al.* Income and racial disparity in household publicly available electric vehicle infrastructure accessibility. *Nat Commun* 15, 5106 (2024). https://doi.org/10.1038/s41467-024-49481-w
12. Gunasekaran, R., Manjunatha, B., Anand, S. *et al.* Prediction of electric vehicle charging demand using enhanced gated recurrent units with RKOA based graph convolutional network. *Discov Appl Sci* 6, 605 (2024). https://doi.org/10.1007/s42452-024-06326-x
13. Rajalakshmi, J., Jeyanthi, S., Bhuvanesh, A. *et al.* Analysis of solar photovoltaic integration and plug-in electric vehicle charging systems with a bidirectional buck-boost DC to DC converter. *J Inst Eng India Ser B* (2024). https://doi.org/10.1007/s40031-024-01185-x
14. Zheng, Y., Keith, D.R., Wang, S. *et al.* Effects of electric vehicle charging stations on the economic vitality of local businesses. *Nat Commun* 15, 7437 (2024). https://doi.org/10.1038/s41467-024-51554-9
15. Hassan, M. Machine learning optimization for hybrid electric vehicle charging in renewable microgrids. *Sci Rep* 14, 13973 (2024). https://doi.org/10.1038/s41598-024-63775-5

Chapter 6

Range and Efficiency

6.1 FACTORS AFFECTING RANGE

The range of an electric vehicle (EV), which is the maximum distance an EV can travel on a fully charged battery, is an essential determinant of an EV's practicality and consumer acceptance. The chapter discusses numerous factors that determine range: technical requirements, environmental factors, and driver behavior. With these factors understood, the people involved in the EV ecosystem can find ways to be efficient and optimize vehicle performance.

The most direct determinant of the EV range is the capacity of its battery pack. Measured in kilowatt-hours (kWh), a battery pack directly affects how much energy an EV can store and release for propulsion; a 50-kWh battery would likely achieve approximately 250 km under standard driving conditions, assuming an efficiency of 20 kWh per 100 kilometers. Energy density, chemistry, and degradation over time vary greatly in differences. The energy density of most EVs' lithium-ion batteries is approximately 250–300 Wh/kg. This forms the basis for a weight-to-range trade-off. Makers [1–3] who want to increase the distance are, in effect, working to increase energy density with new materials and designs that would make possible distances without an increase in weight in proportional sets of the vehicle.

Power electronics, including inverters and controllers, also reduce energy losses during the transformation of electrical energy into mechanical energy. For instance, it is said that the efficiency of Tesla Model 3 permanent magnet synchronous motor is approximately 90%. That is where its competitive range of 400–500 km has come from. In comparison, losses in poor powertrain designs or lack of proper maintenance result in energy loss and, consequently, a low effective range.

The aerodynamic efficiency of the vehicle is very high and associated with the energy consumption of the EV, in terms of the drag coefficient, Cd. Aerodynamic drag, as the name suggests, supplies an opposing force to the vehicle's motion in proportion to the square of its velocity as the vehicle passes through the air. In most highway drives that exceed 100 km/hour, aerodynamic drag is the strong energy consumer. Latest EVs such as the

DOI: 10.1201/9781003608097-6

Lucid Air and Tesla Model S have achieved coefficients of 0.21 and 0.23, respectively. Both these vehicles benefit from streamlined designs and underbody optimization in an attempt to minimize the loss of energy. Conversely, more aerodynamic vehicles with a drag coefficient greater than 0.35, like boxy SUVs, will have less range due to higher air resistance.

The weight of an EV also has several considerations. For example, this is in reference to the weight of the battery pack, the chassis, and the onboard components. The heavyweight of battery packs at 300–600 kg cannot easily optimize the range. For example, a 1,500-kg light-weighted EV will achieve the rate of energy consumption between 15 and 20 kWh per 100 km. However, the rate of energy consumption of a heavyweight luxury SUV of over 2,500 kg would be 25–30 kWh per 100 km. More and more aluminum and carbon fiber-reinforced polymers are used in the construction of EVs to compensate for the weight-induced loss in range. In addition, some energy loss is offset via the adoption of technologies such as regenerative braking, by which kinetic energy gained through deceleration can be recuperated.

Table 6.1 provides a summary of the interconnected relationship among battery capacity, curb weight, and overall driving range for different electric vehicle models.

The battery capacity in a car does not determine that the higher is better because the weight, aerodynamic efficiency, and design of the powertrain vary between two vehicles, as exemplified in Hyundai Kona Electric's higher range compared to Audi e-Tron despite having a smaller battery size because it has lesser weight and higher aerodynamic efficiency.

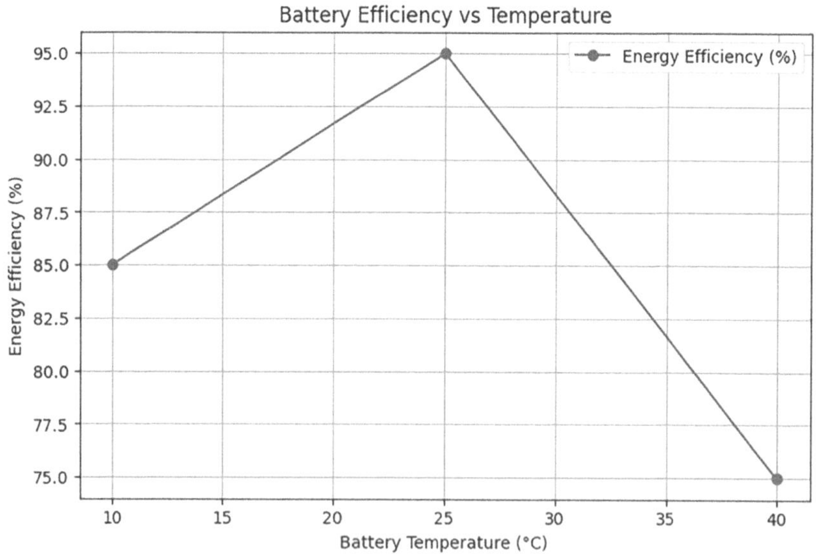

Figure 6.1 Battery Temperature for Energy Analysis.

Table 6.1 Relationship among Battery Capacity, Vehicle Weight, and Range across Different EV Models.

Vehicle Model	*Battery Capacity (kWh)*	*Weight (kg)*	*Range (km)*
Tesla Model 3 standard	50	1,612	423
Nissan Leaf e+	62	1,735	364
Audi e-Tron	95	2,565	359
Hyundai Kona Electric	64	1,685	484

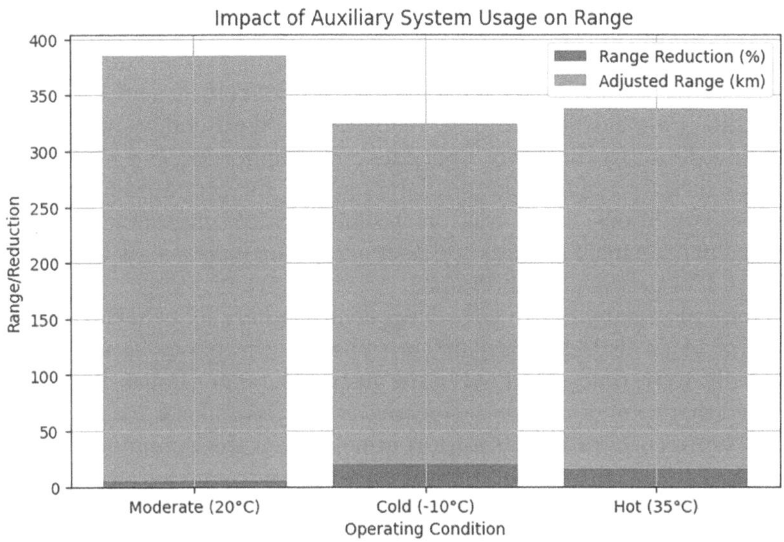

Figure 6.2 Impact of Auxiliary System Use Sets.

Temperature, road conditions, and altitude also affect the range of an EV. The battery is sensitive to temperature. In general, optimal performance occurs in the range 20–30°C. If it goes below freezing, efficiency may fall by as much as 40%, as can be seen from the reduced range of a vehicle under winter conditions. A study conducted by AAA shows that a Tesla Model S lost range from 370 km at 25°C to 220 km at -7°C. This is because of the increased internal resistance in the battery and the energy consumption [4–6] of the heating systems. On the other hand, high temperatures can cause the degradation of the battery and, therefore, have a negative effect on long-term performance.

Road conditions, which include gradient and surface quality, also affect range. Energy consumption increases uphill as the vehicle fights against gravity, while downhill driving allows energy recovery through regenerative braking. Smooth asphalt roads also minimize rolling resistance, thereby allowing efficient energy use. However, rough or uneven surfaces increase resistance, reducing range. Rolling resistance, measured in terms of the coefficient of rolling resistance (CRR), varies between 0.005 for smooth asphalt and 0.02 for rough gravel roads. This will lead to a 10–15% difference in range for the same vehicle under different road conditions.

Table 6.2 provides a detailed analysis of the percentage reductions in range across different operating environments.

The most important factor is driver behavior, which affects EV range. Aggressive acceleration, high-speed driving, and braking frequently increase energy consumption. Generally, the optimal speed range for range efficiency lies in the range of 60–80 km/hour. For example, a Nissan Leaf consuming 15 kWh/100 km at 60 km/hour can increase its consumption to 25 kWh/100 km at over 120 km/h. Similarly, repeated acceleration and deceleration negate the advantage of regenerative braking, making processes inefficient.

Auxiliary systems like heating, ventilation, and air conditioning (HVAC) also contribute to the range loss. The consumption of 2–3 kWh/hour by HVACs directly reduces the available energy for propulsion. In extreme weather conditions, cabin comfort assumes greater importance. Advanced thermal management systems and heat pumps are increasingly being used in modern EVs to reduce such losses.

Range sustainability is finally dependent on charging behavior and management of the battery. Frequent fast charging decreases battery capacity gradually over time and lowers the practical range. It has been recommended that to minimize these negative effects, users keep a state of charge (SOC) between 20% and 80% for routine trips, as this maintains full charge capability for extensive road trips. This allows a smart charging system for optimal pattern adaptation in maintaining the health of batteries while maximizing ranges achieved.

Table 6.3 illustrates that the range of an EV is a complex parameter influenced by technical, environmental, and behavioral variables. Although the range has been constantly improved with battery technology, powertrain efficiency, and aerodynamic design, the environmental and behavioral variables are equally important to consider for the right usage of EVs. Manufacturers

Table 6.2 Analysis of the Impact of Environmental Conditions on the Range of a Mid-Sized EV.

Condition	*Range Reduction (%)*	*Notes*
Sub-zero temperatures	30–40%	Increased battery resistance, heating demand
Hot weather (>35°C)	10–15%	Reduced cooling efficiency, air conditioning
Uphill driving (5% slope)	20–25%	Increased energy demand due to gravity
Rough road surfaces	10–15%	Higher rolling resistance

Table 6.3 Comparison of the Impact of Different Driving and Charging Behaviors on EV Range Sets.

Behavior	*Range Impact (%)*	*Explanation*
Aggressive driving	−20% to −30%	Increased energy demand for acceleration
HVAC use	−10% to −20%	Energy diverted for heating/cooling
Optimal speed driving	+10% to +20%	Reduced aerodynamic and rolling resistance
Frequent fast charging	−5% to −10% (long-term)	Accelerated battery degradation

and users should collaborate in making EVs more efficient and practicable for widespread utilization and sustainability.

6.2 INCREASING EFFICIENCY VIA REGENERATIVE BRAKING

Regenerative braking is one of the most important technologies in EVs that significantly enhanced the efficiency and mileage range of EVs. During deceleration, regenerative braking systems recover energy by converting what would otherwise be lost as heat into electrical energy. The principles, mechanisms, and benefits behind regenerative braking will be explained with the support of quantitative analyses and illustrative data samples.

Therefore, regenerative braking is essentially the reversibility of operation of an electric motor to make it act more like a generator at the application of brakes. If the driver releases the accelerator or even the brake is pressed, electrical energy is first generated and then converted into current in the same electric motor while converting it back toward the battery for recharge. Some variables that are shown to affect regeneration include motor efficiency, battery availability, and manner of braking patterns.

The amount of energy recovered through regenerative braking is proportional to the kinetic energy of the vehicle. Kinetic energy is proportional to the mass of the vehicle and the square of its velocity. For example, a 1,500-kg EV traveling at 60 km/hour has about 208,000 joules of kinetic energy. This is when we are assuming that the regenerative braking system can capture 70% of the energy. Due to this, around 145,600 joules will be converted into electrical energy and stored in the battery. The retrieved energy reduces the total energy of the vehicle, meaning it travels for a longer distance in a single charge.

Table 6.4 shows that the heavier vehicles or higher speed vehicles possess more kinetic energy; hence, their regenerative braking system can recover more energy. But this recovery suffers from efficiency losses in the motor, power electronics, and battery.

Table 6.4 Energy Recovery Potential of Regenerative Braking for Vehicles with Different Weights and Speed.

Vehicle Weight (kg)	*Speed (km/h)*	*Kinetic Energy (J)*	*Recovered Energy (J) (70%)*
1,200	50	115,200	80,640
1,500	60	208,333	145,833
2,000	70	378,472	264,930
2,500	80	617,284	432,099

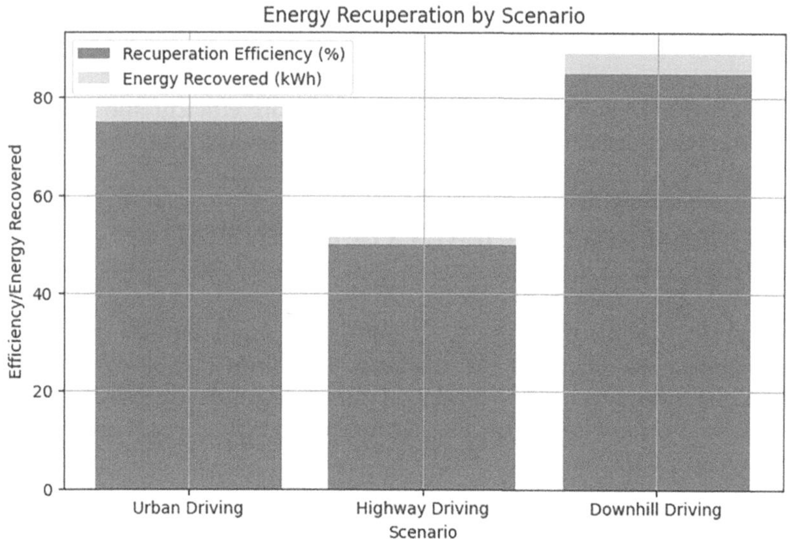

Figure 6.3 Energy Recuperation Scenario Analysis.

The efficiency is not only dictated by the nature of the recovery process but also dependent upon how the recuperated energy has been used within the system. Modern EVs employ sophisticated energy management systems (EMSs) that manage to either prioritize or utilize regenerated power for auxiliary consumers, thus drawing less from the pack and saving more energy. Utilizing advanced algorithms [7–9], the energy produced is distributed intelligently in such a manner that more of the power is used as a means for extending the mileage of these vehicles.

Thermal management of regenerative braking efficiency also plays a significant role. When the kinetic energy is converted into electrical energy, heat generation is applied to the motor and power electronics. This could be a source of thermal loss if the heat generated is not well dissipated. Therefore, the efficiency of cooling systems is critical for maintaining the performance of regenerative braking mechanisms. For example, water-cooled motors dissipate heat much better than the air-cooled ones; thus, the energy recovery efficiency is higher.

Table 6.5 Comparative Analysis of the Energy Recovery Rates and Range Extensions Achieved through Regenerative Braking in Urban versus Highway Driving Scenarios.

Driving Condition	*Energy Recovery Rate (%)*	*Range Extension (%)*
Urban	20–25	10–15
Highway	10–15	5–10

The other role is road conditions and driving patterns. Urban environment or high-frequency stop-and-go activities offer the chances for sustained recovery of energy with regenerative braking systems. For instance, an EV traveling through a city that has lights every 500 m would trigger its regenerative braking numerous times in a kilometer. In contrast, on highways, where brake is applied less, the scope for the recovery of energy is merely limited. Studies have found that on the basis of regenerative braking, 25% of the energy can be saved on urban routes and merely 10–15% on highway conditions.

Iteratively, as per the data in Table 6.5, another aspect that must be considered is how regenerative braking will affect the life of the batteries. Lithium-ion batteries, for instance, tend to shorten their lives due to such repeated cycles of charge and discharge as those brought about by regenerative braking. Advances in battery management systems, however, have mitigated this effect by storing regenerated energy within optimal voltage ranges. For instance, a study on EVs found that with regenerative braking, improved best practices in energy management prolong the life of batteries by as much as 15% because it has an added value of being energy efficient and sustainable.

Manufacturers have also continuously improved in the course of enhancing the effectiveness of the regenerative braking systems. For example, Tesla's current regenerative braking system in Model 3 and Model S is one of the best-integrated braking systems with the vehicle. It allows the drivers to choose various intensities of regenerative braking and therefore provides a customized driving experience with maximum energy recovery. This one-pedal driving is the second feature of e-Pedal in the Leaf, wherein driving the accelerator naturally places the car under regenerative braking, and is easier and efficient.

Therefore, regenerative braking systems improve range while decreasing wear on mechanical braking components. Due to this reason, most deceleration loads are covered by regenerative braking, reducing the rate and frequency of mechanical brake engagements, thereby resulting in low maintenance costs. For example, if an EV uses regenerative braking to provide 70% of its deceleration events, it could ensure that the lifespan of the brake pads lasts up to three times as long as those without regenerative capability.

Table 6.6 Maintenance Cost Savings Associated with Regenerative Braking across Different EV Models, Considering Reduced Brake Pad Replacement Frequency.

EV Model	*Brake Pad Lifespan (km)*	*Cost per Replacement ($)*	*Savings with Regenerative Braking ($)*
Tesla Model 3	100,000	300	600
Nissan Leaf	80,000	250	500
Hyundai Kona Electric	90,000	270	540

As highlighted in Table 6.6, regenerative braking not only enhances efficiency but also delivers tangible economic benefits to EV owners by reducing maintenance expenses.

Future developments are expected in more energy-efficient motors, higher levels of energy storage, and far smarter control algorithms. The smart Advanced Driver-Assistant Systems will enable maximized recovery energy by estimating pre-braking, depending upon road and traffic situations. The ultra-capacitor can recharge as well as deplete with remarkable speed, and it may be used to assist the classic batteries in better exploiting captured regenerative energy.

The concept of regeneration braking is very much clear and suggests energy efficiency as it offers additional mileage along with several other advantages. Regenerative braking saves precious amounts of energy, reduces associated costs of maintenances, and supports efficient transport. On the basis of improvement in EV technology, the advancement of regenerative braking is the most impactful factor for enhancing efficiency and practice for electric transports.

6.3 ENERGY MANAGEMENT SYSTEMS

The heart and soul in optimizing performance, efficiency, and range is the EMSs. The essential constituents—the battery, motor, and auxiliary systems—operate in synchrony through intelligent control of the distribution and utilization of electrical energy. The principles, functions, and developments pertaining to EMSs will be expounded upon by taking into consideration their role in enhancing the range and efficiency.

An EMS should regulate the energy flow within an EV in such a manner that neither wastes it nor consumes it wastefully. The main aim of this EMS includes battery SOC monitoring, energy balancing at drivetrain, and auxiliary loads of climate control system management. Optimal patterns in the use of energy will increase the range of an EV with minimal loss in case the EMS is appropriately designed. For instance, high-order energy

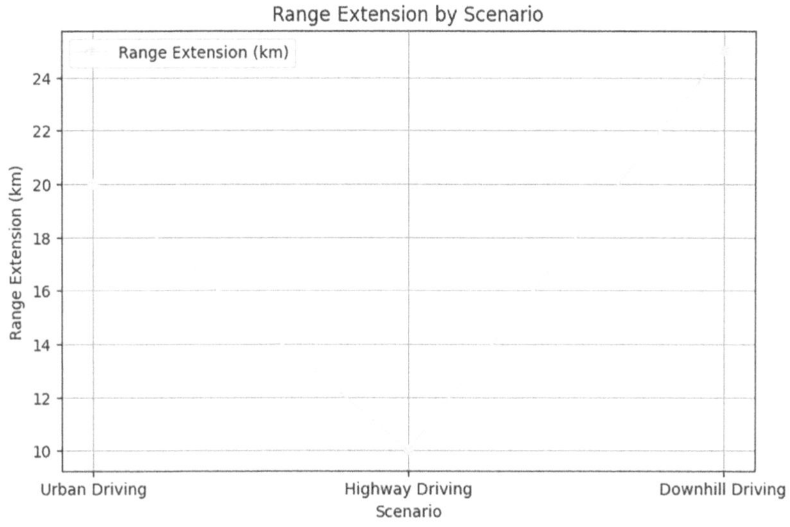

Figure 6.4 Range Extension Scenario Analysis.

Table 6.7 Comparison of Battery Performance under Different Thermal Management Scenarios.

Battery Temperature (°C)	*Energy Efficiency (%)*	*Range (km)*
10	85	340
25	95	380
40	75	300

management algorithms have proven the potential for increasing efficiency throughout a whole vehicle of up to 15%. Consequently, a higher travel range can be achieved whenever the same amount of battery capacity is involved.

Battery management can be cited as one of the main functionalities in an EMS. Because a battery is the main source of energy for an electric car, it is continuously monitored for essential parameters such as voltage, current, temperature, and SOC. Lithium ion, which is commonly used in an EV, should operate at an ideal temperature between 20 and 30°C. An EMS [10–12] prevents thermal runaway of the batteries by activating the cooling or heating solutions to manage the temperature of the batteries. For example, an advanced thermal management system in a Tesla Model 3 car keeps the temperature of the battery at 25°C during high-speed charging, which supports effective energy transference and long-lasting batteries.

As shown in Table 6.7, keeping the battery within the ideal temperature range—around 25°C—maximizes energy efficiency and range. Energy efficiency reduces sharply without good thermal management, particularly at extreme temperatures, thereby adversely affecting the performance of the vehicle overall.

The EMS also optimizes energy distribution to the electric motor, which will propel the vehicle. By controlling the delivery of voltage and current, the EMS ensures that the motor operates at its peak efficiency across different driving conditions. For example, during acceleration, the EMS delivers more current to meet torque demands, while during steady cruising, it reduces energy flow to conserve battery charge. In regenerative braking, the EMS controls the process of converting kinetic energy into electrical energy, returning it to the battery for storage.

Among the issues that involve the EMS, one of the main problems is finding a balance in the energy intake between the drivetrain and the auxiliary systems. Other parts, such as heating, ventilation, and air conditioning, lighting, and infotainment, could be the cause of the excessive energy consumption, especially during extreme climatic conditions. For instance, the use of an HVAC system in freezing temperatures may reduce the range of an EV by at least 20%. To mitigate this, modern EMSs have predictive algorithms and energy-saving modes that adjust auxiliary loads based on driving conditions and user preferences.

The data in Table 6.8 demonstrates why EMS is a necessity in controlling auxiliary loads. An EMS will not waste energy when the weather is tough if the HVAC operation is smartly controlled, thus conserving ranges. The optimization of route and driving behavior is also another crucial application of the EMS. An EMS can predict energy consumption of a given route and suggest optimal routes to the driver by interfacing with onboard sensors and navigation systems. For example, an EMS may advise a driver on the route that consumes less energy if it has lesser inclines, or it might recommend a particular speed to attain a maximum range. In addition, some EMSs use machine learning algorithms that learn about the driver's behavior and tailor their response accordingly, thereby maximizing energy efficiency.

In addition to the above capabilities, energy recovery is another domain where EMSs aid in efficiency. EMSs control regenerative braking such that the maximum possible energy is recovered during braking and downhill driving scenarios. Advanced systems modify the rate of regenerative

Table 6.8 Impact of Auxiliary System Usage on the Range of an EV under Various Operating Conditions.

Operating Condition	*HVAC Usage (kWh)*	*Range Reduction (%)*	*Adjusted Range (km)*
Moderate weather (20°C)	1	5	380
Cold weather (−10°C)	3	20	304
Hot weather (35°C)	2.5	15	323

Table 6.9 Comparison of Energy Recuperation Efficiency under Different Driving Scenarios.

Scenario	*Recuperation Efficiency (%)*	*Energy Recovered (kWh)*	*Range Extension (km)*
Urban driving (frequent stops)	75	3	20
Highway driving (few stops)	50	1.5	10
Downhill driving (steep gradient)	85	4	25

braking dynamically based on battery SOC and vehicle speed. For example, a depleting battery of a vehicle descending a steep hill reduces the intensity of regenerative braking to avoid overcharge, while energy recovery will be maximized in a vehicle with a lower SOC.

Table 6.9 shows that the energy recovered through regenerative braking can extend the range significantly in stopping and steep-hill driving situations. In EMSs, the control strategy has to optimize the regenerative braking process to gain maximum advantage from this scenario. Emerging themes of EMS technology involve integrating the systems with V2G. V2G allows electricity to be exchanged from the grid by the EVs and back to the grid as provided by the EVs at the peak demand periods. This mutual operation is well-managed by an advanced EMS, which includes the operation of running the vehicle with some reasonable amount of energy left for the running of its functions. For example, an EV with a 60-kWh battery can supply up to 10 kWh to the grid without any influence on its driving range by dynamically adjusting to the prevailing SOC and planned usage. Future EMS advances will be made using artificial intelligence (AI) and connected vehicle technologies. AI-enabled EMSs can provide real-time, data-driven predictions from sources including traffic patterns [13–15], weather conditions, and battery health, enabling the ability to predict when energy needs are likely to occur and optimizing the source of energy for the application. Connected vehicle ecosystems allow EMSs to communicate with charging stations and other vehicles to optimize energy usage across a network of EVs.

In summary, EMSs play a key role in optimizing mileage and efficiency in EVs. EMSs can possibly maximize the utility of the stored energy with safe and efficient operation in controlling energy distribution, recuperation, and consumption in a smart and effective manner. The added amplification brought by the integration with advanced technologies and predictive algorithms will then pave the way for more efficient and sustainable electric mobility solutions. With innovation, EMSs are going to maintain their lead in the battle with the energy efficiency challenges of EVs.

REFERENCES

1. An, X., Liu, Y., Liu, Z. *et al.* Research on rapid cycle test method in low-temperature of electric vehicle driving range based on China light-duty vehicle test cycle. *Int J Automot Technol* (2024). https://doi.org/10.1007/s12239-024-00145-0
2. Lin, X., Chen, Z., Zhang, J. *et al.* A multi-objective optimization control strategy of a range-extended electric vehicle for the trip range and road gradient adaption. *Int J Automot Technol* 25, 131–145 (2024). https://doi.org/10.1007/s12239-024-00021-x
3. Singh, A.R., Suresh, K., Parimalasundar, E. *et al.* A high-efficiency poly-input boost DC–DC converter for energy storage and electric vehicle applications. *Sci Rep* 14, 18176 (2024). https://doi.org/10.1038/s41598-024-69254-1
4. Güven, A.F., & Yücel, E. Sustainable energy integration and optimization in microgrids: Enhancing efficiency with electric vehicle charging solutions. *Electr Eng* (2024). https://doi.org/10.1007/s00202-024-02619-x
5. Li, L. Study on optimization of water use efficiency in electric vehicle maintenance process in sustainable green environment. *J Sol-Gel Sci Technol* (2024). https://doi.org/10.1007/s10971-024-06636-x
6. Beigmoradi, S. Harmonizing aerodynamic efficiency, stability, and acoustic excellence: Multi-objective optimization for electric vehicle rear-end design. *Multiscale and Multidiscip Model Exp and Des* 8, 56 (2025). https://doi.org/10.1007/s41939-024-00633-x
7. Vinod, M., & Kishan, D. A dual full-bridge series-series resonant IPT system for ultra-wide-range electric vehicle battery applications. *Electr Eng* (2024). https://doi.org/10.1007/s00202-024-02542-1
8. Edwin Isaac Raj, A., Carmel Sobia, M., & Ravindran, M. Advanced dual-switch converter with proportional integral technique for enhanced efficiency and power quality in electric vehicle energy storage. *Electr Eng* (2024). https://doi.org/10.1007/s00202-024-02807-9
9. Mamo, T., Gopal, R., & Yoseph, B. Modeling and predesign analysis of electric vehicle considering Ethiopian driving cycle. *Int J Automot Technol* (2024). https://doi.org/10.1007/s12239-024-00045-3
10. Kumar, M., Kumar, B., & Rani, A. Flux optimization strategy for efficiency enhancement of induction motor–driven electric vehicle under dynamic operating conditions. *Energy Efficiency* 16, 68 (2023). https://doi.org/10.1007/s12053-023-10149-x
11. Abdullah, M., Moin, L., Ahmed, F. *et al.* Electric vehicle battery temperature control using fuzzy logic. *Aut Control Comp Sci* 58, 237–251 (2024). https://doi.org/10.3103/S0146411624700135
12. Lukuyu, J., Shirley, R., & Taneja, J. Managing grid impacts from increased electric vehicle adoption in African cities. *Sci Rep* 14, 24320 (2024). https://doi.org/10.1038/s41598-024-75039-3
13. Shewalkar, A.G., Dhoble, A.S., & Thawkar, V.P. Review on cooling techniques and analysis methods of an electric vehicle motor. *J Therm Anal Calorim* 149, 5919–5947 (2024). https://doi.org/10.1007/s10973-024-13091-x

14. Gao, C., Zhao, X., Zhu, W. *et al.* Impact load between spherical obstacles and electric vehicle battery packs. *J Mech Sci Technol* 38, 2845–2854 (2024). https://doi.org/10.1007/s12206-024-0504-3
15. Velasquez, J.D.O., Moreno, J.A.G., & Aldana, N.L.D. Experimental platform for studying energy regeneration in electric vehicle powertrains. *J Power Electron* 24, 1751–1765 (2024). https://doi.org/10.1007/s43236-024-00836-y

Chapter 7

Electric Vehicle Maintenance

7.1 BATTERY CARE AND MAINTENANCE

Intricacies in electric vehicle (EV) battery care primarily include charging and temperature management issues, factors that could affect degradation and the life of the battery.

Lithium-ion cells account for the majority of EV batteries, owing to their ability to provide high energy density, longevity, and efficiency. Many hundreds or even thousands of the cells are paired together in the battery pack installed inside an EV for electrical energy dissemination throughout the motor. Typically, the lifetime of an EV battery is estimated in charge–discharge cycles; modern lithium-ion batteries are meant to last between 1,000 and 2,000 cycles, or roughly 8–15 years of normal usage. Proper care can extend that, while neglect or improper handling can significantly shorten it.

Probably the most critical aspect of maintaining an EV battery is proper charging practices. The two main causes of battery degradation are overcharging and deep discharging. The producers advise keeping the state of charge (SOC) between 20 and 80% for daily use. For instance, Tesla advises its clients to charge up to 80% for daily usage and up to 100% if they will be traveling long distances. Deep discharging below 20% SOC results in irreversible damage to the cells, thereby reducing their capacity over time. Constant charging up to 100% would quicken the chemical processes responsible for diminished performance. Finally, temperature also affects the life time of the operation. Currently available EVs [1–3] operate adequately within the ranges of 20 and 40°C. Such frequent occurrences at extreme ends will damage their chemistry. Long-term exposure to temperatures above 50°C results in thermal runaway, where the battery gets extremely overheated without any control and hence gets badly damaged. At low temperatures, below -20°C, the battery capacity gets reduced for some time, resulting in limited mileage and poor performance of the vehicle. Thermal management systems installed in new age EVs monitor and regulate battery temperature using liquid cooling or heating mechanisms.

DOI: 10.1201/9781003608097-7

Table 7.1 provides a detailed comparison of battery performance under varying temperature conditions, illustrating how temperature extremes can impact battery capacity and charging efficiency.

From Table 7.1, it is evident that maintaining an optimal temperature of around 25°C allows the battery to retain 100% capacity and achieve a high charging efficiency of 95%. As the temperature deviates from this range, both capacity retention and charging efficiency decrease significantly. For instance, at –20°C, the capacity retention drops to 60%, while charging efficiency becomes 70%. This brings out the need for parking EVs in shaded garages or areas during hot weather and pre-conditioning systems during colder climates to mitigate the influence of temperature.

Table 7.1 Battery Performance Under Varying Temperature Conditions.

Temperature (°C)	*Capacity Retention (%)*	*Charging Efficiency (%)*
25	100	95
40	95	90
50	85	80
0	80	85
−20	60	70

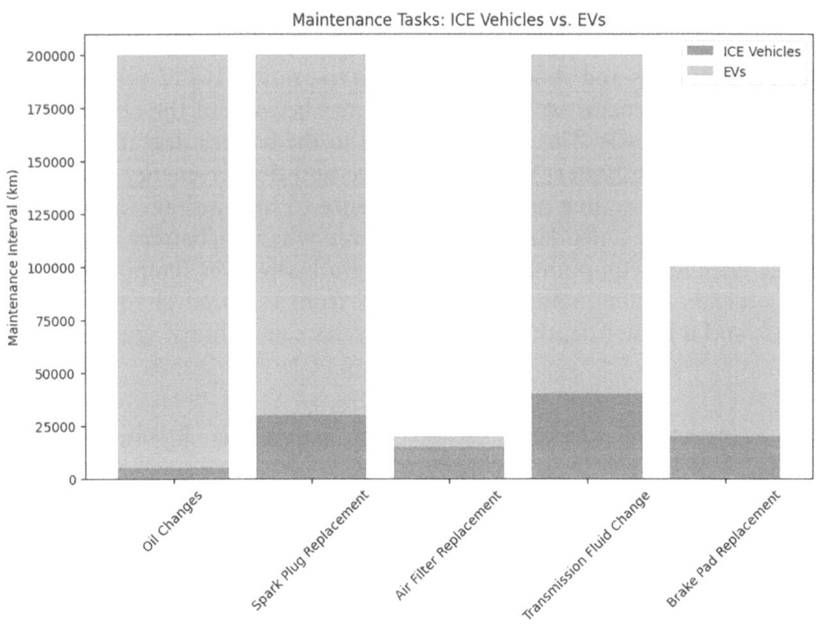

Figure 7.1 Model Maintenance Task Analysis.

Another significant factor that influences the health of the battery is charging speed. It quickly charges, while convenient, and more heat is emitted than during regular charge, which creates some form of thermal stress on battery cells. In spite of these, most of the EVs will cover overheating to protect the battery; however, prolonged exposure can still induce the rapid degradation of batteries. Investigations reveal that vehicles primarily charged using a DC fast charger experience a loss in their capacity of approximately 10% after five years relative to a 5% loss in vehicles that are mostly charged using a Level 2 charger. It empowers the user to make less use of fast charging and make use of slower modes of charging for normal usage.

The role of regenerative braking in maintaining an EV battery is very much unobserved. In the process of regenerative braking, when the vehicle brakes, the kinetic energy generated due to deceleration is converted into electrical energy, which gets stored in the battery. It enhances the energy efficiency and minimizes the usage of traditional braking systems. However, overreliance on regenerative braking leads to more frequent cycling of the battery, thereby increasing wear and tear. Optimizing the balance between regenerative braking and conventional braking [4–6] optimizes the battery health. Table 7.2 provides an overview of how charging practices and regenerative braking influence battery degradation rates, focusing on balanced usage patterns.

As shown in Table 7.2, the balanced use of Level 2 charging and regenerative braking will minimize the annual degradation rate to between 1.5% and 1.8%, thereby prolonging the lifespan to around 14–15 years. High-frequency DC fast charging and over-regenerative braking can speed up the degradation rates and shorten the lifespan to around 10–12 years.

The natural process of degradation occurs because of the chemical reactions inside the cells. The active materials in the battery degrade over time, which causes the battery's ability to store and deliver energy to decrease. Calendar aging, cycling aging, and exposure to high voltage contribute to this degradation. Calendar aging occurs even when the battery is not in use, as the chemical components deteriorate gradually over temporal instance. The other is cycling aging, which comes from repeated charge–discharge cycles, and a higher depth-of-discharge cycles causes more aggressive wear.

Table 7.2 Impact of Charging Practices and Regenerative Braking on Battery Degradation.

Charging/Braking Method	*Annual Degradation Rate (%)*	*Expected Lifespan (Years)*
Level 2 charging	1.5	15
Frequent DC fast charging	3.0	10
Balanced regenerative braking	1.8	14
Excessive regenerative braking	2.5	12

Higher voltage exposure, especially if the battery is kept close to 100% SOC for a longer time, accelerates the growth of lithium plating, which in turn reduces the capacity.

Monitoring and maintenance can neutralize battery degradation. New EVs are fitted with battery management systems (BMSs) that are monitoring parameters [7–9], including voltage, current, and temperature levels. This way, safe operation is guaranteed by balancing the cells within the pack to avoid overcharging and deep discharging. Software updates coming from the manufacturer can further improve the BMS function in relation to better health management of the battery.

In addition to the software approaches, physical examination of the battery pack is another necessary requirement. Cleaning the terminals and removing corrosion should be included in every electrical connection checkup. Higher resistance by corrosion may translate into a larger amount of energy lost and even overheating. It is also essential that one checks his battery pack physically from time to time for physical damage such as swelling or leaking as such situations could compromise safety and efficiency. Table 7.3 shows the maintenance schedule for EVs, showing general guidelines for recommended maintenance at different mileage intervals.

Following the maintenance schedule in Table 7.3 will allow users to anticipate any potential problems and, thus, extend the life of the battery and achieve optimal performance. In conclusion, preventing the degradation of EV batteries would necessitate proper charging practice, temperature management, balanced usage patterns, and regular inspections. Following these principles and using leading-edge battery management technology, EV users can drive the batteries hard, promising to last them the life of their vehicle with as much efficiency and reliability as is possible. Continuous research and development are addressing the battery degradation, enabling a surge in EV adoption as consumer confidence increases regarding the long-term reliability of electric mobility.

Table 7.3 Recommended Maintenance Schedule for EV Batteries.

Maintenance Activity	*Frequency*	*Purpose*
SOC calibration	Every 6 months	Ensures accurate charge levels
Thermal management system check	Annually	Verifies temperature control
Battery terminal cleaning	Every 12 months	Prevents corrosion
Software update	As provided by original equipment manufacturer (OEM)	Enhances BMS functionality
Physical inspection for damage	Annually	Ensures safety and reliability

7.2 TIRE MAINTENANCE

Tire maintenance represents a critical consideration toward ensuring safety, overall operating efficiency, and driving range of EVs. Unlike an internal combustion engine vehicle (ICEV), specific factors define EVs, changing the way tires experience wear and therefore the requirements to service the tires. These include higher body weights by virtue of battery packs and immediate torque production. The regenerative braking system makes all the differences concerning the life of an EV and the performance of tires. All these factors will be discussed elaborately in the following section, showing how proper tire care and maintenance schedule and tire choices make a significant difference in one's driving experiences.

One of the reasons why EV tires have a greater maintenance problem is increased weight from heavy battery packs. The weight can be as heavy as hundreds of kilograms, which would definitely add more weight compared to its counterpart, ICEV. For instance, Tesla Model 3 weighs about 1,700 kg, about 300 kg heavier than an ICE sedan with a comparable size. The added weight will increase the load on tires and cause faster wear and higher rolling resistance. Proper tire pressure is necessary to manage this added load and ensure even wear. Underinflated tires result in excessive rolling resistance, which will reduce energy efficiency and increase tire wear. Overinflated tires can cause reduced contact with the road, compromising traction and safety. Table 7.4 provides a summary of the relationship among

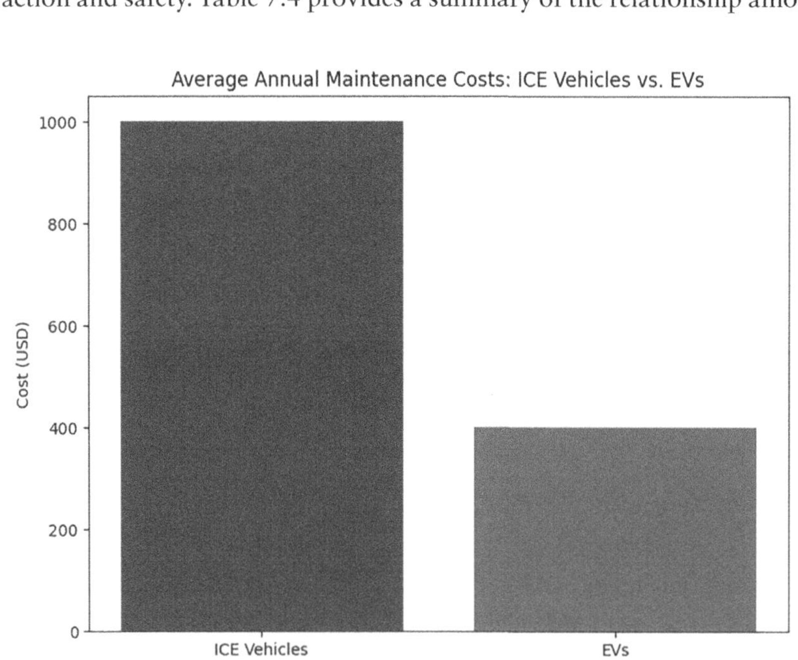

Figure 7.2 Model Average Maintenance Cost Analysis.

Table 7.4 Impact of Tire Pressure on Rolling Resistance and Energy Consumption.

Tire Pressure (psi)	*Rolling Resistance Coefficient*	*Energy Consumption (kWh/100 km)*
30	0.015	18.5
35	0.012	17.0
40	0.010	16.2
45	0.009	15.8
50	0.008	15.5

tire pressure, rolling resistance, and energy consumption in EVs as elaborated. From the data in Table 7.4, maintaining the optimal tire pressure is important for safety and efficiency.

From Table 7.4, it is evident that if the tire pressure is maintained at around 40–45 psi, rolling resistance is reduced and a saving of around 15.8 kWh per 100 km can be achieved. However, if the pressure is decreased to 30 psi, then the coefficient for the rolling resistance increases to 0.015, and hence more energy is consumed, that is, 18.5 kWh per 100 km. This is the reason why regular checking and refilling the correct pressure into the tires are critical to avoid energy loss and minimize extra tire stress.

The second characteristic in EV tire care is instant torque. Electric motors offer maximum torque just about instantaneously, while the torque of an ICEV is delivered gradually as the RPM of the engine increases. For example, the Porsche Taycan [10–12] goes from 0 to 60 mph in 2.4 seconds. Its instant torque delivery is certainly useful for acceleration but does mean the tires endure a bit of added stress at each sudden start or aggressive launch. This could cause uneven wear, particularly on rear tires of rear-wheel-drive EVs or on front tires of front-wheel-drive EVs over temporal instance.

Regenerative braking is the third hallmark characteristic of EVs that contributes to tire maintenance. In regenerative braking, kinetic energy from braking is captured and then turned into electrical energy to charge the battery. The braking forces of regenerative braking are indeed less taxing on the traditional brake components but do impact the tires—the front tires moreso, in cases where the system is regenerative front-wheel drive. The absence of this natural effect might mean that the asymmetrical wear of tires causes one to wear out at the front axle more than those at the back, necessitating a call for rotation at periodic intervals. Tire rotation distributes wear evenly on all tires, hence prolonging their life. Table 7.5 presents the summary of tread wear rates as observed on EVs with and without tire rotations.

As presented in Table 7.5, the front tires wear at 1.5 mm per 10,000 km without tire rotation, while the rear tires wear at 0.8 mm per 10,000 km. Regular tire rotation will make all tires wear around 1.0 mm per 10,000 km. It not only prolongs the tire life but also stabilizes and balances the vehicle better.

Table 7.5 Tread Wear Rates in EVs with and without Tire Rotation.

Tire Position	*Without Tire Rotation (mm/10,000 km)*	*With Tire Rotation (mm/10,000 km)*
Front	1.5	1.0
Rear	0.8	1.0

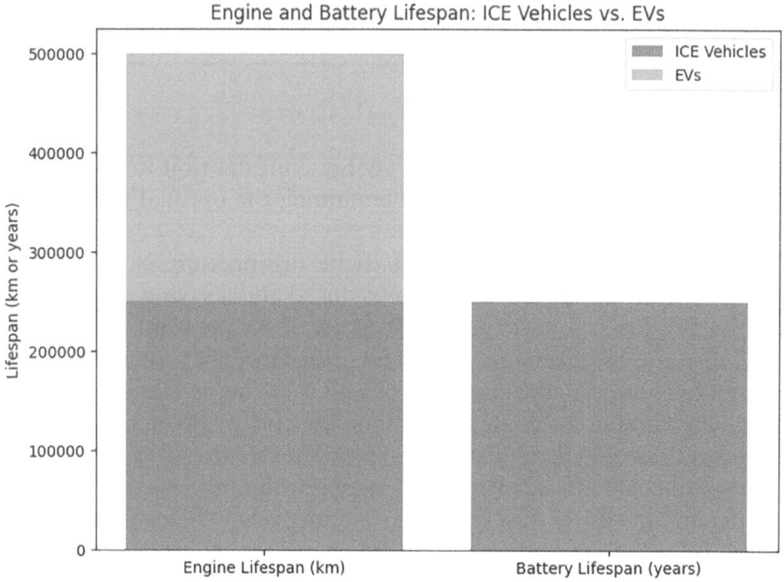

Figure 7.3 Model's Energy and Battery Analysis.

The next important part of EV tire maintenance is tire alignment. Misaligned wheels lead to uneven tire wear, loss of energy efficiency, and compromise safety. EVs are more sensitive to alignment problems because of their higher weight and the precise handling characteristics. A few degrees of misalignment can lead to significant wear on one side of the tire, requiring it to be replaced before its time. Regular checking of tire alignment, especially after a hit on a pothole or curb, will maintain proper performance.

Tire type is also an important aspect of EV maintenance. Most manufacturers of EVs recommend or provide low-rolling-resistance tires to minimize energy loss and maximize the range. Specialized compounds and tread patterns reduce rolling resistance, which are often implemented to ensure there is enough traction and durability for the tires. However, low-rolling-resistance tires may wear out faster than conventional tires under aggressive driving conditions. Balancing performance and durability requires careful consideration of the driving style and typical road conditions. Table 7.6 compares the performance characteristics of low-rolling-resistance tires and conventional tires in EV applications, highlighting their advantages and limitations.

Table 7.6 Performance Characteristics of Low-Rolling-Resistance Tires Versus Conventional Tires.

Characteristic	*Low-Rolling-Resistance Tires*	*Conventional Tires*
Rolling resistance coefficient	0.010	0.015
Energy efficiency (kWh/100 km)	16.0	17.5
Tread wear life (km)	40,000	50,000
Traction performance	Moderate	High

From Table 7.6, it can be seen that low-rolling-resistance tires are more energy efficient, so they reduce energy usage to 16.0 kWh per 100 km compared with the conventional tire energy usage at 17.5 kWh. However, the tread wear life is less for 40,000 km compared to 50,000 km in conventional tires. Traction performance is also relatively lower and, therefore, not the best for high-performance or all-weather uses. In this process, one should consider keeping the above trade-offs in mind when choosing tires for their EVs.

The other important aspect of tire maintenance is proper storage of unused tires. Tires must be stored in a cool and dry place free from direct sun and all sources of heat, which may degrade the rubber compound. Tires that are not mounted on rims should be placed upright to avoid deformation, and once these tires have been in use, they should be clean and free of oil, grease, and all forms of contaminants to enhance their lifespans.

In conclusion, tire maintenance for EVs requires a multilayered approach considering electric propulsion, weight of the vehicle, and its driving dynamics. Many of these habitual practices—proper tire pressure, tire rotations, proper alignment, and proper choice of tires—are able to ensure long life of tires and the satisfactory performance of any vehicle. By adhering to these guidelines, comprehending how tire characteristics will meet the demands of electric vehicles will enhance safe, efficient, and economical driving. Improvements in the technologies used for designing tires and material science are continually expected as technologies advance for EVs.

7.3 COMMON EV ISSUES AND TROUBLESHOOTING

EVs are highly advanced machines that provide efficient, clean, and reliable transportation. However, like all mechanical and electrical systems, EVs are prone to some of the following problems that may happen due to wear and tear, software glitches, or other reasons. It is essential to be aware of such common issues and their respective troubleshooting strategies for maintaining the performance and longevity of EVs. This section presents the most recurrent issues reported for EVs and their root cause, along with the best practices for the mitigation of these issues.

Degradation in driving range is the most frequently reported issue with EVs. Driving range depends on numerous factors: state of health of the battery, weather conditions, driving habits, and auxiliary systems such as heating and air conditioning. For instance, a new EV, rated at 300 km range, might lose as much as 10% of its range after 5 years of constant use. In most cases, such loss in range is usually attributed to battery degradation because the lithium-ion cells lose their capacity due to repeated cycles of charging and discharging. This is limited by keeping the optimal SOC at 20% and 80%. Keeping off extreme temperatures that boost the aging of batteries is also helpful.

The second problem facing most EV owners is the charging issue. This encompasses slow charging, not starting to charge, and interruptions in charging. Slow charging results when one uses a lower capacity charger or has poor electrical connections. For instance, it would take a minimum of 40 hours to fill the battery [13–15] of Tesla Model 3 if a Level 1 charger at 120 volts is used, but it can save around 8 hours of charging time when a Level 2 charger at 240 volts is used. Charging could fail due to faults in a broken charging cable or connector, some software issues with the BMS within an EV, etc. Troubleshooting such problems will require checking on the charging equipment, verification of the connections, and upgrading of software packages in the car. Table 7.7 presents typical charging times for some of the most common EV models at various levels and thus explains why one has to ensure that the proper charging infrastructure is used during the process.

As can be seen in Table 7.7, the charging time is significantly reduced by using a Level 2 charger or DC fast charger. Thus, owning an EV makes the process much more user-friendly. Fast chargers should not be used frequently because more frequent use of fast chargers will degrade the batteries. EV software faults affect several systems, such as infotainment interfaces, navigation, and driving controls. For example, some of the common issues faced by some of the owners of Tesla cars are malfunctions in the working of the touch screen. For example, glitches cause an inability to use climate and navigation features in the car. The company rectifies all these errors through software updates. As the EV software is upgraded, errors are rectified, and

Table 7.7 Charging Times for EVs Using Different Charging Levels.

EV Model	*Level 1 Charger (120V)*	*Level 2 Charger (240V)*	*DC Fast Charger (50 kW)*
Tesla Model 3	40 hours	8 hours	30 minutes
Nissan Leaf	35 hours	7.5 hours	40 minutes
Hyundai Kona	38 hours	9 hours	45 minutes
Chevrolet Bolt	36 hours	8.5 hours	35 minutes

Table 7.8 Common Issues in Regenerative Braking Systems and Solutions.

Issue	*Possible Cause*	*Recommended Solution*
Reduced energy recovery	Faulty sensors or worn components	Replace sensors or brake pads
Uneven braking	Misaligned brake components	Adjust brake alignment
Regenerative braking failure	Software error	Update system software

systems would work better, along with additional features. Another component that often shows problems is regenerative braking systems. One of the most important features of EVs is regenerative braking, which captures energy during deceleration and stores it in the battery. Malfunctions of regenerative braking often occur as less than adequate energy recovery, erratic braking, or total failure of the system. Most problems result from faulty sensors, worn parts, or incorrect software. In diagnosing and repairing the regenerative braking system, one should check on the sensors, brake pads, and electronic control systems. Table 7.8 presents the list of some of the common regenerative braking issues and their probable solutions.

Table 7.8 clearly shows that regenerative braking problems require both hardware modifications and software updates. Problems like these should be identified early on by regular maintenance and diagnostic tests so that they do not create further complications.

Temperature-related challenges majorly affect EV owners. Extreme coldness temporarily impairs battery performance, consuming more energy; extreme heat causes the battery to degrade faster, leading to lower charging efficiency. In the example case, an electric car running under extremely low temperature can lose its maximum range to about 30% because not only does it need to keep its cabin warm, but the temperature of its battery also has to be kept high. Pretreatment of the battery and the cabin before setting out, parking in shaded or insulated places, and deploying thermal management systems enclosed in the vehicles are part of the troubleshooting about temperature-related problems. Noise and vibration problems are not as prevalent in EVs compared to that in ICEVs but may occur due to issues in drivetrain or suspension components. Unusual sounds related to acceleration or braking might be due to electric motors, gearboxes, or wheel bearings. They can also include misaligned wheels, worn parts of the suspension, or irregular wear on tires. The diagnosis involves inspecting the damaged parts and repairing or replacing the affected parts. For example, Table 7.9 presents some common causes of noise and vibration in EVs and suggested remedies. As seen in Table 7.9, identification and prompt resolution of noise and vibration problems are crucial to have a smooth and safe drive. Regular check-up and maintenance would prevent this from happening.

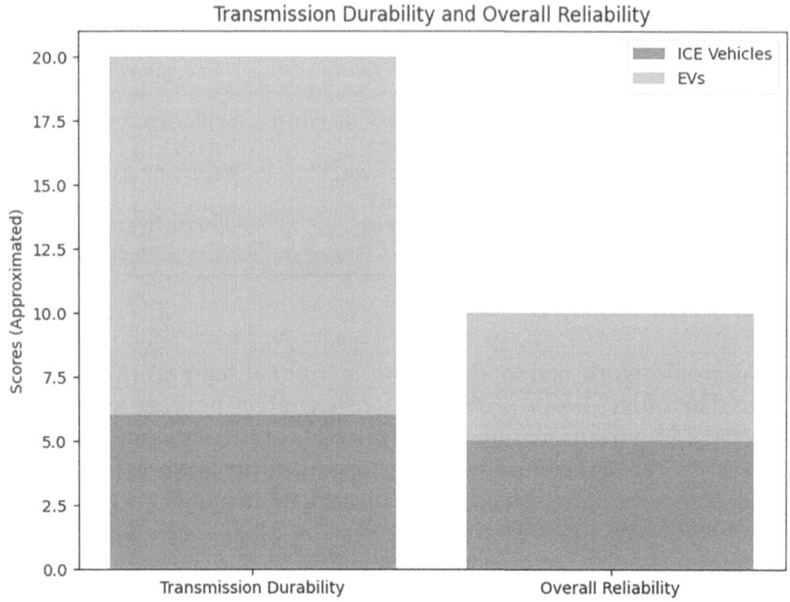

Figure 7.4 Model's Transmission Durability Analysis.

Table 7.9 Noise and Vibration Issues in EVs and Remedies.

Symptom	*Possible Cause*	*Remedy*
Noise during acceleration	Motor or gearbox issue	Inspect and repair motor/gearbox
Vibration while driving	Misaligned wheels	Perform wheel alignment
Noise during braking	Worn brake components	Replace brake pads or rotors

In conclusion, although EVs are engineered for reliability and require less maintenance than their counterpart, ICEVs, issues cannot be avoided with them as well. Some of the common issues include range reduction, failure in charging, malfunctioning software, failure in regenerative braking, temperature-related problems, and noise or vibration problems. All these can be overcome with proper troubleshooting and regular maintenance. If one understands the problems and their solutions, EV owners will be able to ensure that the longevity, safety, and efficiency of their vehicles are maintained. Advancements in EV technology and diagnostics will further reduce the challenges mentioned above and make electric mobility much smoother. The maintenance needs of an EV differ very much from that of an ICEV.

7.4 MAINTENANCE COMPARED TO INTERNAL COMBUSTION ENGINE (ICE) VEHICLES

EVs have been praised for their very low maintenance requirements, but their comparison with ICEVs shall uncover some specific sets of challenges and benefits of each type. The following section explains the differences in practices, costs, and reliability of the component by EVs and ICEVs to set a base understanding of the implications of ownership and long-term use. One of the major differences between EVs and ICEVs is that EVs do not have an internal combustion engine. ICEVs rely on complex engines with many moving parts, including pistons, valves, crankshafts, and timing belts. These components must be serviced regularly, including oil changes, spark plug replacements, and timing belt inspections, to keep them running smoothly. On the other hand, EVs use electric motors, which are much less complicated and have significantly fewer moving parts. For example, a conventional ICEV engine may contain over 1,000 moving parts, whereas an EV motor contains fewer than 20. Such simplicity minimizes the possibility of mechanical failures and also does away with the routine maintenance task of oil changes. Table 7.10 presents the key maintenance tasks of ICEVs compared with EVs, highlighting the maintenance complexity of EVs over ICEVs.

As illustrated in Table 7.10, some of the maintenance activities applicable to ICEVs do not apply to EVs due to the absence of components such as internal combustion engines and conventional transmissions. For example, frequent oil change applicable to ICEVs is eliminated in EVs and thus saves much time and money for the owner.

The maintenance activity for brake systems is also quite different between the two types of vehicles. ICEVs use conventional friction-based brakes, which degrade much quicker and require replacing brake pads and rotors more often than the regenerative type of braking in an EV, which turns some of the kinetic energy into electrical energy to charge the battery through the motor to slow the car. This implies that the old brakes are rarely used, and the brake pads are used for a longer time cycle. According to various studies,

Table 7.10 Maintenance Tasks for ICEVs Versus EVs.

Maintenance Task	*ICEVs*	*EVs*
Oil changes	Every 5,000–10,000 km	Not applicable
Spark plug replacement	Every 30,000–50,000 km	Not applicable
Air filter replacement	Every 15,000–20,000 km	Cabin air filter only
Transmission fluid change	Every 40,000–60,000 km	Not applicable (no gearbox)
Brake pad replacement	Every 20,000–30,000 km	Extended interval (regenerative braking)

Table 7.11 Average Annual Maintenance Costs for ICEVs Versus EVs.

Vehicle Type	*Average Annual Maintenance Cost (USD)*
ICEVs	\$800–\$1,200
EVs	\$300–\$500

it has been indicated that the EV brake pads last up to 100,000 km, and for the ICEVs, it is up to 30,000 km, thus saving maintenance.

The second major difference lies in the cooling systems. An ICEV requires extensive cooling systems since it produces much heat from combustion. These may include radiators, coolant reservoirs, and hoses, which would need regular inspections and coolant replenishment. In contrast, EVs do not produce much heat and tend to require much simpler thermal management systems to keep the battery and motor at acceptable temperatures. This reduces complexity, leading to fewer maintenance requirements and reduced component failure. The average annual maintenance costs for ICEVs and EVs are summarized in Table 7.11 to compare the maintenance cost difference.

Table 7.11 shows that EVs incur much lower maintenance cost compared to ICEVs. The reasons are primarily due to the absence of oil change, fewer brake replacements, and use of simpler cooling systems. For instance, an EV owner can save from \$5,000 to \$7,000 in maintenance cost over a period of 10 years than an ICEV owner.

However, there are also unique maintenance problems in EVs. The most critical component of an EV is its battery pack. This has a very big share of the overall price. Modern lithium-ion batteries, used in today's EVs, are built for 8–15 years and need proper maintenance to prevent battery degradation early in life. There are also possibilities of degradation related to deep discharge, overcharging, and even extreme temperatures. Replacement of batteries often costs between \$5,000 and \$15,000, depending on the type of vehicle. It is for this reason that the owners must most definitely follow some best practices regarding battery care.

Software maintenance is another essential factor in owning an EV. An EV depends much more on software than its ICE equivalents for operations such as energy management, autonomous driving, and infotainment. Therefore, it is essential to provide the requisite software updates to achieve the best performance, along with maximum security, and to rectify the bugs, add novel features, and reduce energy consumption. Software maintenance is less labor intensive than maintenance in mechanical types but requires a stable Internet connection and occasional attention from the owner.

Table 7.12 summarizes the long-term reliability and durability difference between an ICEV and an EV, which explains the factors influencing the maintenance strategy.

Table 7.12 Long-Term Reliability and Durability of ICEVs Versus EVs.

Attribute	*ICEVs*	*EVs*
Engine lifespan	200,000–300,000 km	Not applicable (electric motor lifespan > 500,000 km)
Battery lifespan	Not applicable	8–15 years
Transmission durability	Frequent repairs required	Minimal issues (single-speed)
Overall reliability	Moderate	High

Table 7.12 shows that, on the whole, EVs are far more reliable than their ICE counterpart over a period of time. An electric motor with a minimal number of moving parts is far longer lasting than an internal combustion engine, which requires a lot to maintain the wear and tear. The conventional transmission part that makes up for a good percentage of failures of ICE-based vehicles does not exist in an EV.

The environmental implications of maintenance activities are also a factor that separates EVs and ICEVs. Used engine oil, transmission fluid, and air filters from ICEVs become waste products, which then must be disposed of in order not to contribute to environmental pollution. EVs produce fewer waste products, although lithium-ion battery recycling is their biggest concern. Better battery recycling technologies are expected to minimize the environmental impacts of EVs further.

In conclusion, the maintenance requirements for EVs are markedly different from those of ICEVs, and EVs offer significant advantages in terms of simplicity, cost savings, and reliability. While ICEVs require frequent attention to engine, transmission, and braking systems, EVs benefit from fewer moving parts, regenerative braking, and simplified thermal management systems. However, the special care of the battery and software updating requires proactive maintenance for EVs. Identifying these differences and integrating the technological advancements in designing EVs will help realize more sustainable and cost-effective vehicles. As the adoption of EVs continues to grow, the maintenance landscape is expected to evolve further, with advancements in diagnostics, repair practices, and battery recycling driving the transition to a cleaner and more efficient automotive future.

REFERENCES

1. Li, L. Study on optimization of water use efficiency in electric vehicle maintenance process in sustainable green environment. *J Sol-Gel Sci Technol* (2024). https://doi.org/10.1007/s10971-024-06636-x
2. Görçün, Ö.F., Simic, V., Kundu, P. *et al.* Electric vehicle selection for industrial users using an interval valued intuitionistic fuzzy COPRAS-based model. *Environ Dev Sustain* (2024). https://doi.org/10.1007/s10668-024-05562-w

3. Horesh, N., Trinko, D.A., & Quinn, J.C. Comparing costs and climate impacts of various electric vehicle charging systems across the United States. *Nat Commun* 15, 4680 (2024). https://doi.org/10.1038/s41467-024-49157-5
4. Polat, A.O., Erden, B.C., Kul, S. *et al.* Light electric vehicle performance with digital twin technology: A comparison of motor types. *Arab J Sci Eng* 49, 7209–7222 (2024). https://doi.org/10.1007/s13369-023-08668-x
5. Zhu, T., Boyles, S.D., & Unnikrishnan, A. Battery electric vehicle traveling salesman problem with drone. *Netw Spat Econ* 24, 49–97 (2024). https://doi.org/10.1007/s11067-023-09607-x
6. Xia, W., Luo, C., Cai, L. *et al.* Environmental impact study of the sightseeing electric vehicle supply chain based on the B2C e-commerce model and LCA framework. *Energy Inform* 7, 138 (2024). https://doi.org/10.1186/s42162-024-00446-9
7. Agbesi, P.K., Ruffino, R., & Hakovirta, M. Creating an optimal electric vehicle ecosystem: An investigation of electric vehicle stakeholders and ecosystem trends in the US. *SN Bus Econ* 4, 28 (2024). https://doi.org/10.1007/s43546-024-00624-7
8. Dhairiyasamy, R., Gabiriel, D., Bunpheng, W. *et al.* A comprehensive analysis of India's electric vehicle battery supply chain: Barriers and solutions. *Discov Sustain* 5, 361 (2024). https://doi.org/10.1007/s43621-024-00595-7
9. Kek, V., Kandasamy, J., Vezhavendhan, R. *et al.* Digital technology adoption model for electric vehicle battery recycling supply chain—an influential relationship mapping. *Oper Manag Res* 17, 1469–1508 (2024). https://doi.org/10.1007/s12063-024-00515-y
10. Filyushov, Y.P., & Filyushov, V.Y. Increasing the mileage of an electric vehicle without changing the capacity of the battery. *Russ Electr Engin* 95, 553–558 (2024). https://doi.org/10.3103/S1068371224700627
11. Saravanan, K., Sivasubramanian, M., Gopinath, N.P. *et al.* A 31 L multilevel inverter topology with less switching devices for hybrid electric vehicle applications. *Sci Rep* 14, 27459 (2024). https://doi.org/10.1038/s41598-024-78529-6
12. Kumar, G., & Altalbe, A. Artificial intelligence (AI) advancements for transportation security: In-depth insights into electric and aerial vehicle systems. *Environ Dev Sustain* (2024). https://doi.org/10.1007/s10668-024-04790-4
13. Huang, B., Yan, H., & Zhou, J. Research on dispatchable potential calculation and energy joint optimization of electric vehicle clusters under the background of carbon neutralization. *Clean Techn Environ Policy* (2024). https://doi.org/10.1007/s10098-024-03068-7
14. Pinto-Bautista, S., Baumann, M., & Weil, M. Prospective life cycle assessment of an electric vehicle equipped with a model magnesium battery. *Energ Sustain Soc* 14, 44 (2024). https://doi.org/10.1186/s13705-024-00475-y
15. Kim, S., Yoo, Y., Choi, M. *et al.* The effect of dynamic wireless charging systems on electric vehicle activation and sustainability. *Clean Techn Environ Policy* (2024). https://doi.org/10.1007/s10098-024-03043-2

Chapter 8

Environmental Impact

8.1 CARBON EMISSIONS AND ENERGY SOURCES

Electric vehicles (EVs) are one of the possible solutions to the pressing challenge of carbon emissions since these emissions cause most of the global climatic changes. The next section unravels the relationship among EVs, carbon emissions, and energy sources by focusing on how EVs could reduce greenhouse gas (GHG) emissions if coupled with sustainable energy systems.

The transportation sector is one of the world's major carbon dioxide (CO_2) emitters, accounting for around 23% of energy-related CO_2 emissions. The most significant contributors [1–3] are internal combustion engine vehicles (ICEVs), emitting around 4.6 metric tons of CO_2 per vehicle annually. The emissions come not only from tailpipe outputs but also from the fuel production and refining processes, collectively referred to as "well-to-wheel" emissions. The EVs, by contrast, produce zero tailpipe emissions, making it an attractive option for reducing the carbon footprint related to transportation.

The carbon impact of EVs is heavily influenced by the type of electricity that is used for recharging. Electricity production varies significantly around the world, including sources such as coal and natural gas, along with renewables including wind, solar, and hydroelectric power. The intensity of emissions from electricity generation is measured in grams of CO_2 per kilowatt-hour (gCO_2/kWh). For example, coal-based power plants average at 820 gCO_2/kWh, while solar and wind energy have virtually no emissions, less than 20 gCO_2/kWh, for the whole life cycle.

Such a difference explains why the choice of energy sources is a key factor in ascertaining the actual environmental benefits of EVs. To understand better the relationship between electricity sources and EV emissions, Table 8.1 provides a comparison of average lifecycle emissions for different electricity generation technologies. Lifecycle emissions encompass emissions from construction, operation, and decommissioning of power plants, providing an all-rounded view of their environmental impact.

From Table 8.1, renewable energy sources, such as wind, solar, and hydropower, have dramatically lower emissions than fossil fuels. EVs have much higher lifecycle emissions than hybrids when charged with electricity

DOI: 10.1201/9781003608097-8

Table 8.1 Lifecycle Emissions of Electricity Generation Technologies.

Energy Source	*Lifecycle Emissions* (gCO_2 / kWh)
Coal	820
Natural gas	490
Nuclear	12
Solar photovoltaic (PV)	20
Wind	15
Hydropower	24

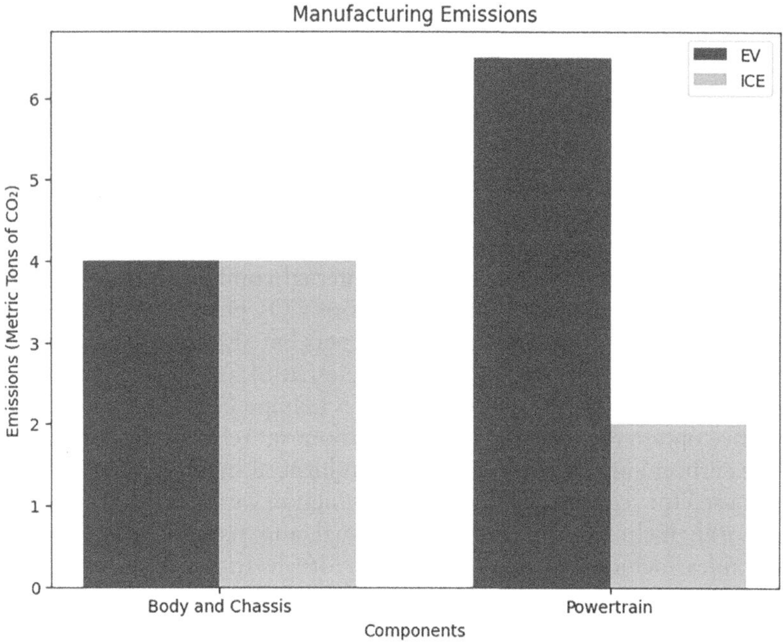

Figure 8.1 Emissions Analysis.

produced from coal. However, if renewable penetration is sufficiently high, EVs exhibit huge emission savings compared to their internal combustion engine (ICE) counterparts. For instance, in Norway, whose electricity supply exceeds 98% hydropower, the total emissions of a set of average ICEVs is less than 5% comprised by EVs.

Another important aspect is the regional variation of electrical grids. Table 8.2 shows the average emissions intensity of electricity generation in various countries. From the table, it is evident that regional policies and

Table 8.2 Average Emissions Intensity of Electricity Generation by Country (gCO_2 / kWh).

Country	*Emissions Intensity (gCO_2 / kWh)*
Norway	10
Germany	310
United States	430
India	720
China	650

the compositions of electricity grids play a crucial role in the environmental advantage of EVs.

From Table 8.2, it is clear that the emissions intensity in countries like Norway is orders of magnitude lower than that in nations such as India and China, where coal dominates the energy mix. This variation implies that the same EV model may have vastly different environmental impacts depending on where it is used. This would require a transformation of the grid into renewable power in high-emission intensity countries for full utilization of the reduction in carbon footprints during the process.

The other area is the production and recycling of the battery, which forms a vital part of an EV's operation and also contributes to the lifecycle emissions of EVs. The energy [4–6] from this process is costly, accounting from an estimated 56–494 kg of CO_2 per kWh of battery capacity. For instance, in a car with an average battery capacity of 60 kWh, the CO_2 emission for the production phase sums up to about 3.36 metric tons. This calls for the need of manufacturing in cleaner ways and the addition of renewable energy in the battery production facility. To give an unbiased view of emissions, there is a need to compare the lifecycle emissions of EVs and ICEVs. Table 8.3 summarizes total lifecycle emissions, which include manufacturing, operation, and end-of-life processes, for a mid-sized passenger vehicle.

Table 8.3 shows that the lifecycle emissions for EVs charged by renewable energy sources are drastically less compared to that of the ICEs. Hence, the

Table 8.3 Lifecycle Emissions of Mid-Sized Passenger Vehicles (Metric Tons of CO_2).

Vehicle Type	*Manufacturing*	*Operation (150,000 km)*	*End-of-Life*	*Total Emissions*
ICEV	7	23	1	31
EV (coal-powered)	9	21	0.5	30.5
EV (renewable energy-powered)	9	1	0.5	10.5

overall emission reduces to approximately two-thirds when it comes to total emission. On the contrary, the differences shrink drastically as well, considering that EVs were charged using electricity generated through coal-based power plants.

EVs are the future of automobiles and hold the potential to drastically reduce carbon emissions if transition to low-carbon electricity sources is supported. Policymakers should, therefore, focus on renewable energy integration, grid modernization, and battery recycling so that maximum environmental advantage is extracted from EVs. In an integrated model of industries and governments, EVs can play a vital role in fulfilling global climate goals and sustainably shaping a future for transportations.

8.2 LIFECYCLE ASSESSMENT OF ELECTRIC VEHICLES

The lifecycle assessment (LCA) is one of the most important methodologies carried out to analyze the environmental impact of a given product or system throughout its whole life cycle—from extracting raw materials up to the final disposal or recycling of the last product. LCA for EVs gives a comprehensive understanding of the full environmental footprint, taking

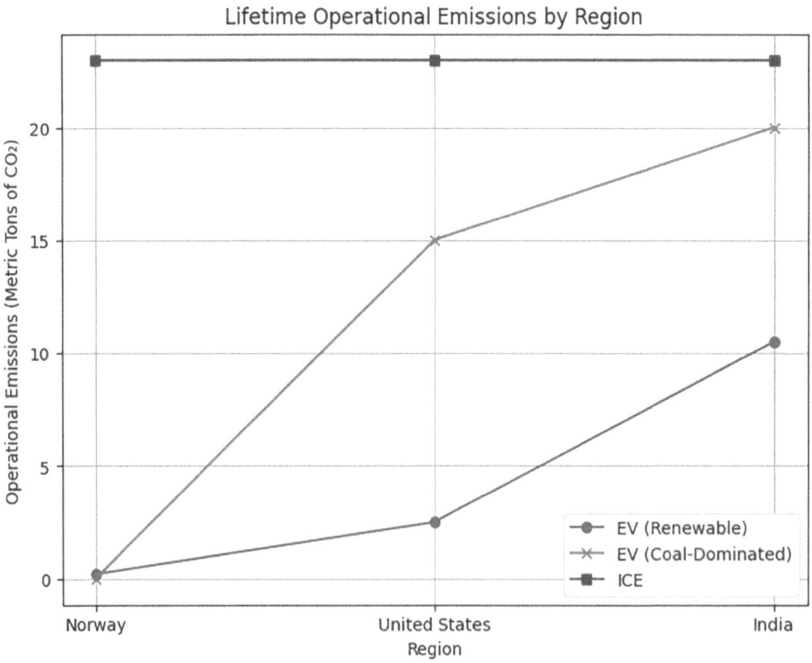

Figure 8.2 Lifespan Analysis of EVs.

into consideration each stage of involvement: manufacturing, operation, and end-of-life. This section further breaks down the LCA of EVs and the resultant environmental implications compared to those of ICEVs and, consequently, explores an avenue to achieve increased sustainability.

A car's LCA would then begin at its raw materials phase. As already indicated, it is dominated primarily by its battery system. Its core constituents include lithium, cobalt, nickel, and manganese. Lithium-ion batteries comprise more than half the market of existing EVs in use. These materials are energy-intensive to extract, which leads to significant environmental impacts that range from habitat destruction to water pollution and even GHG emissions. For instance, it requires approximately 2.3 kg of CO_2 to produce 1 kg of lithium carbonate, an essential component in battery productions. Cobalt mining, however, is usually associated [7–9] with high energy usage and social challenges, and as such, it remains one of the key areas where sustainable improvements are needed for EV supply chains. The largest carbon-intensive part of the life cycle of EVs remains its manufacturing. A significant amount of emissions can be traced to the source of the battery. Table 8.4 outlines that as follows.

As shown in Table 8.4, EV manufacturing generates higher GHG emissions than ICE manufacturing primarily due to the battery. For example, a 60 kWh battery produces about 6.5 metric tons of CO_2. This makes the industry focus on using cleaner energy sources for the manufacturing facilities and improving the technologies used in the batteries for the reduction of these emissions. At the same time, a better environmental impact minimization process needs to be used in the operational phase for the EVs in different scenarios.

Most considerable environmental benefits will be observed in the operational phase of EVs. In contrast to the ICEV, which burns fossil fuels and emits CO_2, the EV provides zero tailpipe emissions. GHG emissions in this phase depend entirely on the electricity mix used for charging. The operational emissions of EVs are zero in regions where the penetration of renewable energy is very high, like Norway, which sources almost 100% of its electricity from hydropower. In coal-dominated grids, like India, the operational emissions of EVs can be equivalent to or even higher than those of efficient ICEVs. Table 8.5 compares the lifetime operational emissions of EVs and ICEVs across different electricity grids.

Table 8.4 GHG Emissions from Vehicle Manufacturing (Metric Tons of CO_2).

Vehicle Component	*EV Manufacturing*	*ICE Manufacturing*
Battery production	6.5	0
Vehicle body and chassis	4.0	4.0
Total	10.5	4.0

Table 8.5 Lifetime Operational Emissions of EVs and ICEVs (Metric Tons of CO_2).

Region	*EV (Renewable Source-Powered)*	*EV (Coal-Powered)*	*ICEV*
Norway	0.2	-	23
United States	2.5	15	23
India	10.5	20	23

Table 8.5 shows that EVs driven by renewable sources have a dramatic decrease in lifetime operational emissions in comparison with ICEVs. Operational emissions of an EV are almost negligible in Norway, whereas in the United States, those will be much lower than those of ICEVs. However, in India, as coal is the dominant source in terms of energy mix, the benefits from EVs are reduced; hence, there is a necessity for an energy transition that reduces more pollutants to see the true results of EVs.

Another very important process of the LCA of EVs is the end-of-life phase. Recycling and disposing of EV batteries is necessary at the end-of-life stage to reduce all the impacts on the environment and to recycle valuable materials. Lithium-ion battery recycling extracts lithium, cobalt, and other

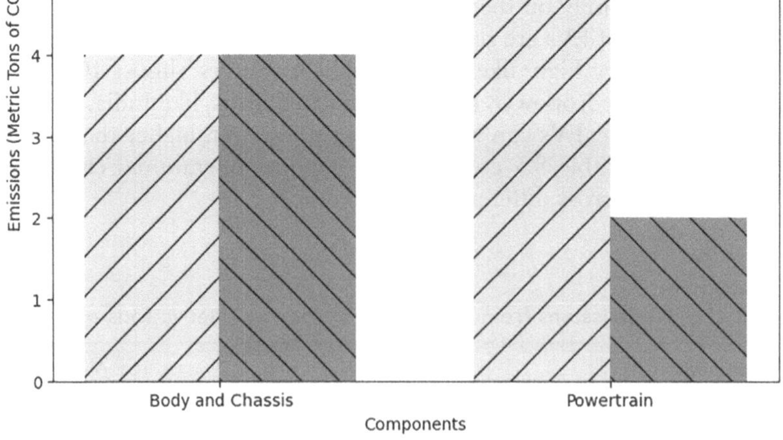

Figure 8.3 Manufacturing Emission Analysis.

Table 8.6 Emissions from Battery Recycling and Disposal (kg CO_2 per Battery).

Recycling/Disposal Method	*Emissions (kg CO_2)*	*Materials Recovered (%)*
Direct recycling	50	80
Pyrometallurgical recycling	200	60
Disposal in landfills	0	0

materials, reducing the demand for virgin raw materials and minimizing the overall footprint of all EV production. Although recycling [10–12] is an energy-intensive and energy-emitting process, Table 8.6 shows the comparison of emissions between the two after recycling and disposal.

Table 8.6 shows that direct recycling of batteries offers both emissions and material recovery balance by producing only 50 kg CO_2 per battery with the recovery of 80% material. Pyrometallurgical recycling is very energy-intensive and low in efficiency but manages to recover around 60% material. Disposal through landfills seems to be completely emission-free at least in the short run, but huge environmental and resource losses are in fact incurred here, as valuable material is not retrieved and may, for instance, leach into waterways.

When the total lifecycle emissions of EVs are compared with those of ICEs, EVs generally have an overall lower footprint, provided their electricity comes from a low-carbon grid. In contrast, operational emissions are somewhat offset by savings in emissions through renewable charging. Poor performance is also associated with challenges in the extraction of raw materials for these batteries and challenges in their production and end-of-life management for optimal environmental friendliness.

This means, consequently, that LCA for EVs involves the involvement of different, sometimes conflicting, stages of its cycle, leading to an overall negative impact on its environment. A drastic reduction of GHG emissions is possible with a linked approach focusing on the decarbonization of electric grids, more improved technology within batteries, and a proper sustainable strategy at the end-of-life cycle. Policy-making and industrial efforts must focus on such sectors in an effort to scale the environmental potential of EVs as the driving force of sustainability within transport infrastructure.

8.3 COMPARING ENVIRONMENTAL FOOTPRINTS

The environmental footprint of a car is described as the complete impact of making, using, and dismantling it through parameters like GHG emission, resource usage, and energy consumption. The comparison of the life cycle of EVs and ICEVs allows for a detailed analysis of every step of the life cycle of both in relation to their respective environmental footprint. These impacts are assessed, and areas where improvement is required regarding sustainability are highlighted.

Both EVs and ICEVs have a great impact at the production stage. There is, however, a difference in both processes regarding carbon intensity. For instance, the production of EVs requires more energy consumption and raw material usage, especially when using large lithium-ion batteries. The batteries consume much of the emissions. For instance, the manufacture of a 60-kWh lithium-ion battery will emit about 6.5 metric tons of CO_2. On the other hand, generating a powertrain for an ICEV, which includes the engine and the fuel system, will produce approximately 2 metric tons of CO_2. Table 8.7 shows the total manufacturing emissions of EVs and ICEVs.

As shown in Table 8.7, the overall manufacturing emissions for EVs are much higher than those for ICEVs mainly because of the battery production process. However, these upfront emissions are often offset during the operational phase, where EVs have a lower environmental footprint due to their reliance on electricity rather than on fossil fuels.

The operational phase of vehicles is the most important determinant of their environmental impact. ICEVs emit GHGs directly from the combustion of fossil fuels, and a mid-sized gasoline vehicle will emit around 23 metric tons of CO_2 in its lifetime, which is about 150,000 kilometers. On the other hand, although EVs create no tailpipe emissions, they produce their own unique form of tailpipe emissions based on the electrical grid mix. Given that a very large number of regions utilize mainly coal-based energies within their mixes, the emissions from EVs can easily resemble or exceed that of ICEVs in those places. Table 8.8 presents the lifetime [13–15] process emission of EVs and ICEVs by region.

As shown in Table 8.8, EV operational emissions rely heavily on the electric power grid. With close to 100% electricity sourced from hydropower in Norway, EVs significantly reduce their lifetime emissions when compared to a similar ICE process. For the United States, a mixed grid would yield moderate operating emissions. For India where coal is the main power source, the differential benefits of EVs would not be as evident. This explains why decarbonization of grids is an important part of how much environmental value EVs will bring.

Resource consumption is another crucial environmental footprint of both types of vehicles. ICEVs depend on finite fossil fuels, which are extracted

Table 8.7 Total Manufacturing Emissions for EVs and ICEVs (Metric Tons of CO_2).

Vehicle Type	*Emissions from Body and Chassis*	*Emissions from Powertrain*	*Total Manufacturing Emissions*
EV	4.0	6.5	10.5
ICEV	4.0	2.0	6.0

Table 8.8 Lifetime Operational Emissions by Region (Metric Tons of CO_2).

Region	*EV (Renewable Energy-Powered)*	*EV (Coal-Dominated Grid)*	*ICEV*
Norway	0.2	-	23
United States	2.5	15	23
India	10.5	20	23

Table 8.9 Resource Consumption and Recycling Potential for EVs and ICEVs.

Vehicle Type	*Primary Resource Consumption (kg)*	*Recycling Potential (%)*
EV	Lithium (10 kg), cobalt (15 kg), nickel (25 kg)	70–80
ICEV	Oil (10,000 liters over lifetime)	1

from the ground, refined, and transported at vast environmental cost. The other is in the material needs for EVs such as lithium, cobalt, and nickel, which are used in battery production. Although these materials are recyclable, the extraction process comes with significant environmental and social costs such as habitat destruction and water pollution. Recycling processes for lithium-ion batteries may recover up to 80% of materials, thus minimizing new mining operations. Table 8.9 compares the resource consumption and recycling potential of EVs and ICEVs.

From Table 8.9 it can be inferred that EVs could recover much of the recycled material compared with its counterpart (ICEVs), which could burn great amounts of oil that can be recycled. Due to this reason, there is an urgent developmental need to establish efficient recycling technologies as a means to mitigate the adverse effects of electric vehicles on the sustainability of life.

Recycling involves disposal, which has challenges of environmental pollution and waste management. For an ICEV, for instance, the engine oils and coolants need to be disposed of as they consist of hazardous materials with possible risks to the environment. while EVs have their batteries as a critical component by which toxic substances such as lithium and cobalt can be released if not handled properly. Nevertheless, advances in the recycling of batteries and the application of second-life use of spent batteries in energy storage may alleviate these negative effects in the future.

This section goes into great detail about the complexity of how these vehicles impact the environment. This includes covering all aspects, such as carbon emissions, LCAs, and comparative environmental footprints between EVs and ICEVs. The key takeaways are summarized with extra data integrated to understand this issue more deeply.

Table 8.10 Lifetime Operational Emissions of EVs and ICEVs (Metric Tons of CO_2).

Region	*EV (Renewable Grid)*	*EV (Coal Grid)*	*ICEV*
Norway	0.3	-	23
United States	3.5	16	23
India	12	22	23

Electric Vehicles and Carbon Footprint Reduction

With an immense scope to reduce carbon emissions, critical contributors to global climate change, EVs emit no direct tailpipe emissions, while ICEVs rely on the burning of fossil fuels. Thus, the real environmental benefits depend on the source of electricity for charging. In areas dominated by renewable sources of energy like Norway, operational emissions from EVs are low. In a coal-dominated grid like India, the benefits are less.

For instance, Table 8.10 outlines CO_2 emissions associated with the lifecycle operation of EVs and ICEVs, and presents a comparison of these emissions in different electricity grids that cover 150,000 km of travel.

With respect to striking regional variations in emissions, shifting to cleaner electricity grids should remain paramount to deliver environmental benefits through EVs.

Lifecycle Assessment of Electric Vehicles

LCAs provide a holistic view of the environmental impacts of EVs from production through operation and disposal. The production phase for EVs is considered to be more resource-intensive compared to that of ICEVs primarily because of manufacturing lithium-ion batteries. It uses a 60-kWh battery that equates to roughly around 6.5 metric tons of CO_2, while for the same operation, powertrains on traditional ICEVs produce only 2 metric tons of CO_2. Nevertheless, this preliminary implication is balanced by operational sets as no tailpipe emission happens with EVs.

Table 8.11 summarizes the lifecycle emissions for EVs and ICEVs, offering a comparative view of their environmental impacts.

Table 8.11 shows that EVs charged on renewable grids show much lower lifecycle impacts compared to ICEVs. But when they are charged on coal-heavy grids, their lifecycle emissions can be greater than those of ICEVs, indicating the need for decarbonizing energy systems.

Resource Use and Recycling Potential

The environmental impacts of EVs also go beyond emissions to resource consumption. Lithium, cobalt, and nickel are critical materials used in EV

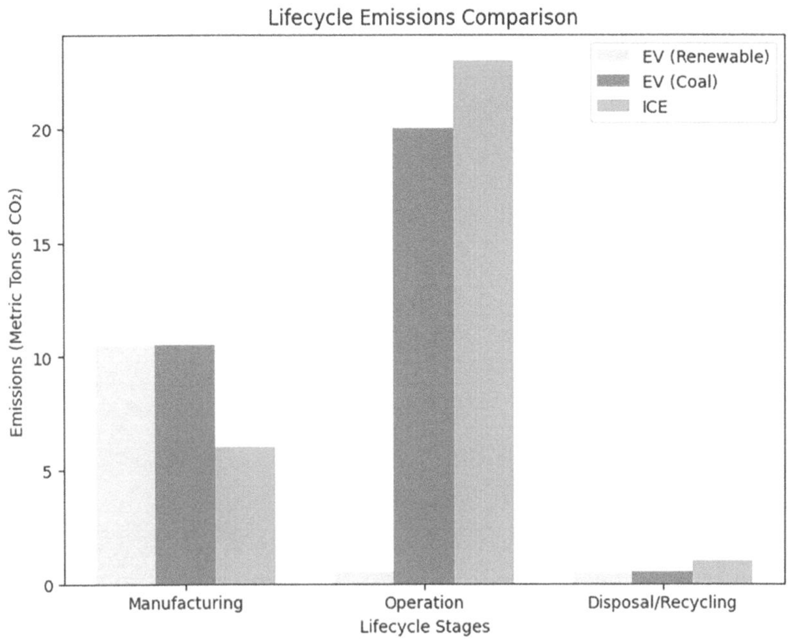

Figure 8.4 Lifecycle Emission Analysis.

Table 8.11 Lifecycle Emissions of EVs and ICEVs (Metric Tons of CO_2).

Stage	*EV (Renewable Grid)*	*EV (Coal Grid)*	*ICEV*
Manufacturing	10.5	10.5	6
Operation	0.5	20	23
Disposal/recycling	0.5	0.5	1
Total	**11.5**	**31**	**30**

batteries, and their extraction has substantial environmental and social implications. For example, extracting 1 kg of lithium carbonate emits approximately 2.3 kg of CO_2 . On the other hand, ICEVs are highly dependent on crude oil; an average mid-sized vehicle consumes around 10,000 liters over its lifetime.

Recovery from end-of-life battery recycling contributes toward reducing the reliance on virgin sources. Using direct recycling techniques, 80% of used-battery material may be recovered in addition to achieving sustainable benefits with the development of EVs. Table 8.12 presents the comparison between the consumption of natural resources and the possibility of recycling EVs versus ICEVs.

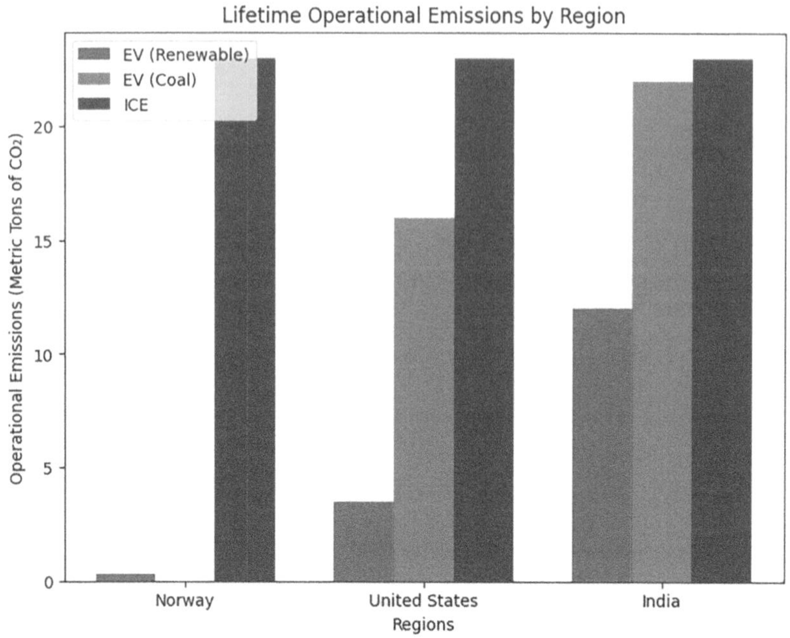

Figure 8.5 Lifetime Operational Emission Analysis.

Table 8.12 Resource Consumption and Recycling Potential for EVs and ICEVs.

Resource	*EV Usage (kg)*	*ICE Usage (Liters)*	*Recycling Potential (%)*
Lithium	10	-	70–80
Cobalt	15	-	60–70
Nickel	25	-	50–60
Crude oil	-	10,000	0

As shown in Table 8.12, as much as the EVs consume extractable resources, their recyclability far outreaches that of ICEVs because of the use of nonrecyclable fossil fuels.

Comparison of Environmental Footprints

Comparison of the environmental footprints of EVs and ICEVs reveals the advantages and challenges faced by each system. EVs have high operational efficiencies; hence, they have zero tailpipe emissions, while their manufacturing processes raise many issues. On the other hand, ICEVs have low emissions from manufacturing but have a high impact due to the operational emissions that are continually caused by fossil fuels.

Newer technologies in the production of EVs, including solid-state batteries, are developed to minimize the emissions of the manufacturing process and maximize energy efficiency. Likewise, further reductions in operational emissions for EVs will come from advances in renewable energy integration and smart grids.

The environmental impact of EVs is a complex interplay of their production, operation, and disposal phases. Although EVs have a good potential for greatly reducing carbon footprints, that still hinges upon the state of the electricity grid, advancement in battery technology, and recycling processes. The future of sustainable transportation requires a systemic approach that integrates renewable energy, efficient recycling methods, and technological innovation as alternatives for ICEVs. Hence, this will lead to the road toward a more sustainable, cleaner transportation paradigm for these EVs.

In conclusion, a comparison of the environmental footprints of EVs and ICEVs shows that although manufacturing emissions for EVs are substantially greater, there is a considerable advantage from the operating phase. Regions with a clean electricity grid further reinforce the point that this process will have fewer resources and more recycling potential in the benefits toward sustainability from EVs. Nevertheless, it would take a whole holistic approach for EVs to reach their full potential for such benefits, including the electrification of the grid, further battery development, and recycling. When these sorts of constraints are relieved, then it is safe to say that EVs can play an important role in mitigating the transport environment's impact and transitioning into a sustainable future.

REFERENCES

1. Xia, W., Luo, C., Cai, L. *et al.* Environmental impact study of the sightseeing electric vehicle supply chain based on the B2C e-commerce model and LCA framework. *Energy Inform* 7, 138 (2024). https://doi.org/10.1186/s42162-024-00446-9
2. Mustafa, S., Shi, Y., Adan, D.E. *et al.* Role of environmental awareness & self-identification expressiveness in electric-vehicle adoption. *Transportation* (2024). https://doi.org/10.1007/s11116-024-10515-3
3. Lyu, W., Hu, Y., Liu, J. *et al.* Impact of battery electric vehicle usage on air quality in three Chinese first-tier cities. *Sci Rep* 14, 21 (2024). https://doi.org/10.1038/s41598-023-50745-6
4. Chang, S.Y., Huang, J., Chaveste, M.R. *et al.* Electric vehicle fleet penetration helps address inequalities in air quality and improves environmental justice. *Commun Earth Environ* 4, 135 (2023). https://doi.org/10.1038/s43247-023-00799-1
5. Pirmana, V., Alisjahbana, A.S., Yusuf, A.A. *et al.* Economic and environmental impact of electric vehicles production in Indonesia. *Clean Techn Environ Policy* 25, 1871–1885 (2023). https://doi.org/10.1007/s10098-023-02475-6

6. Baumgartner, C., & Helmers, E. Life cycle assessment of electric kick scooters: Consolidating environmental impact quantification and concluding climate-friendly use options. *Environ Sci Eur* 36, 96 (2024). https://doi.org/10.1186/s12302-024-00920-x
7. Zeng, T., Zhang, H., Moura, S.J. *et al.* Economic and environmental benefits of automated electric vehicle ride-hailing services in New York city. *Sci Rep* 14, 4180 (2024). https://doi.org/10.1038/s41598-024-54495-x
8. Zorlu, A., & Ozcan, M. Techno-economic and environmental assessment of a solar-powered electric vehicle charging station in Türkiye. *Arab J Sci Eng* (2024). https://doi.org/10.1007/s13369-024-09868-9
9. Li, K., Shi, L., Zhang, Y. *et al.* Global selection appraisal study for heat pump system of electric vehicle based on energetic, economic, and environmental analysis. *Carb Neutrality* 3, 16 (2024). https://doi.org/10.1007/s43979-024-00091-x
10. Abdullah, M., Moin, L., Ahmed, F. *et al.* Electric vehicle battery temperature control using fuzzy logic. *Aut Control Comp Sci* 58, 237–251 (2024). https://doi.org/10.3103/S0146411624700135
11. Wang, G., Wang, L., Wang, L. *et al.* Electric vehicle charging load modeling based on influence factor analysis. *J Shanghai Jiaotong Univ (Sci)* (2023). https://doi.org/10.1007/s12204-023-2663-2
12. Pinto-Bautista, S., Baumann, M., & Weil, M. Prospective life cycle assessment of an electric vehicle equipped with a model magnesium battery. *Energ Sustain Soc* 14, 44 (2024). https://doi.org/10.1186/s13705-024-00475-y
13. Dalkic-Melek, G., Saltik, E.C., & Tuydes-Yaman, H. Electric vehicle (EV) market penetration in countries with rising motorization rates. *Int J Civ Eng* (2024). https://doi.org/10.1007/s40999-024-01039-z
14. Görçün, Ö.F., Simic, V., Kundu, P. *et al.* Electric vehicle selection for industrial users using an interval-valued intuitionistic fuzzy COPRAS-based model. *Environ Dev Sustain* (2024). https://doi.org/10.1007/s10668-024-05562-w
15. An, X., Liu, Y., Liu, Z. *et al.* Research on rapid cycle test method in low-temperature of electric vehicle driving range based on China light-duty vehicle test cycle. *Int J Automot Technol* (2024). https://doi.org/10.1007/s12239-024-00145-0

Chapter 9

Market Trends and Adoption

9.1 GLOBAL MARKET TRENDS IN ELECTRIC VEHICLES

The electric vehicle (EV) market has experienced an unprecedented growth over the last decade. Technological advancements, supportive government policies, environmental concerns, and changing consumer preferences are some of the drivers of this phenomenon. This chapter looks into the trends defining the EV market, particularly in terms of growth patterns, regional dynamics, and emerging challenges.

Growth Patterns of the EV Market

The global EV market had been growing by a compound annual growth rate (CAGR) of nearly 25% from 2015 to 2023. In 2022, there was a sales of around 10 million units of EVs on a global platform; the number was around 2.2 million units as in 2019. Such growth tracks indicate consumer acceptability and have a healthy supply chain that upholds the viability of production and deployments. Interestingly, electric cars accounted for around 14% of all new car sales globally in 2022, a number that increased from just 4% in 2020.

One of the main driving factors to that growth has been the falling battery pack costs for EVs. Over the period of 2010–2023, the price for lithium-ion battery packs reduced by an average from $1,200/kWh to below $150/kWh. It has thereby brought EVs closer in cost to competing with ICE-powered vehicles in several markets, since upfront costs were much reduced.

Regional Market Trends

Regional adoption of EVs is not uniform since the policies at the local levels, infrastructural conditions, and consumer attitudes are quite different. The maximum demand for EVs was noticed in the Asia-Pacific region with China at its head, with over 50% of total sales across the world in the year 2022. Europe and North America follow the above, with Europe reporting excellent growth mainly because of stricter emission regulations and policies from governments. For the United States [1–3], the sales of EVs

DOI: 10.1201/9781003608097-9

Table 9.1 Regional Share of EV Sales in 2022.

Region	*EV Sales (Units, Millions)*	*Share of Global Market (%)*
Asia-Pacific	5.2	52
Europe	3.1	31
North America	1.6	16
Rest of the world	0.1	1

skyrocketed in 2022 to about 800,000 units with an annual increment of 65%. India and Southeast Asia are emerging markets that remain in the early stages of adopting EVs and stand at a penetration rate of below 2% in 2023. But going forward, they are likely to see a lot of growth with strong government support and investments in charging infrastructure.

Table 9.1 shows that the Asia-Pacific region has been leading the world in EV sales. China has promoted electrification quite aggressively and hence is the most dominant region. Europe has a very strong regulatory environment, focusing on reducing emissions and supporting EVs. North America, mainly the United States, is quickly catching up but EV sales is restricted in some regions due to the lack of charging infrastructure.

9.2 INFLUENCE OF GOVERNMENT POLICIES

Government policies around the world are the most important drivers for the adoption of EVs. In Norway, for example, EVs accounted for 86% of new car sales in 2022, where tax incentives, free tolls, and abundant charging networks were provided. A target of achieving 100% zero-emission vehicle sales in the European Union (EU) by 2035 has added to the boom. China's New Energy Vehicle (NEV) mandate, which requires manufacturers to produce a percentage of their sales as EVs, keeps driving growth in the domestic market.

The United States, besides this, has also been offering incentives toward the purchase of EVs. Under the Inflation Reduction Act of 2022, particular models were eligible for tax credits up to a maximum amount of $7,500. Results [4–6] are going to be much faster than those envisioned in previous studies. Research has revealed that 40% of new car sales in the United States will come in the form of EVs by 2030.

Charging Infrastructure Advancement

The EV market expansion relies on charging infrastructure. The count of public charging stations worldwide increased from 1 million in 2018 to over 3.5 million in 2023. This new infrastructure relieved the anxiety about running out of range for developed markets.

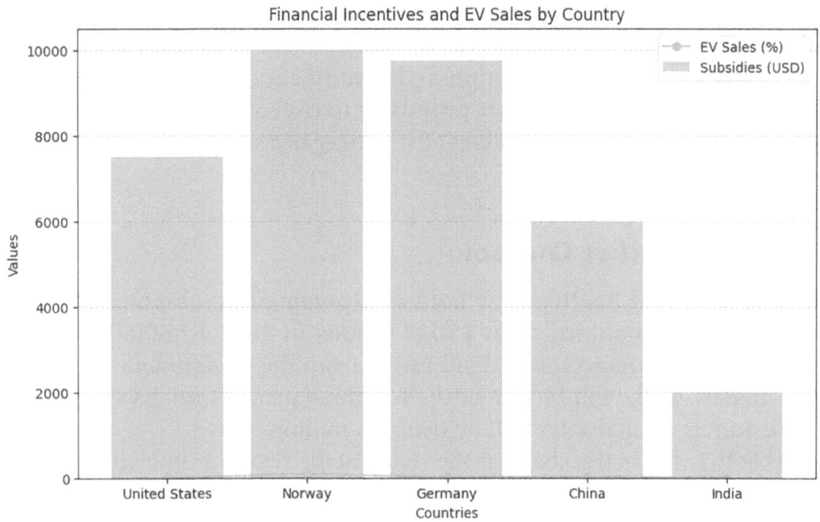

Figure 9.1 Financial Incentive Analysis.

Table 9.2 Growth in Public Charging Stations (2018–2023).

Year	*Number of Public Charging Stations (Millions)*
2018	1.0
2019	1.4
2020	1.8
2021	2.4
2022	3.0
2023	3.5

China has the largest public charging infrastructure in the world: it had around 1.8 million of such public charging points in 2023. Europe has the second largest public charging infrastructure, and Germany, the Netherlands, and France are heavy investors in the ultra-fast network of charging infrastructures. National Electric Vehicle Infrastructure (NEVI) in the United States will have to use $5 billion to erect a nationwide network process.

Table 9.2 shows the steady growth in charging infrastructure over the last 5 years with an annual growth rate of around 25%. This trend is reflective of increased investments by both governments and private players to support the growing EV fleets.

Challenges and Opportunities

The rapid growth of the EV market brings its own set of challenges such as high upfront costs, rural regions having little or no charging infrastructure, and bottlenecks in the supply chains for even the most critical minerals: lithium and cobalt. Conversely, however, these are the challenges that provide opportunities for innovation and investments. Technologies for solid-state batteries seem particularly promising to overcome fears of the cost and range of batteries, while recycling efforts may be the way out of reliance on scarce minerals.

Emerging Market Outlook

India, Brazil, and South Africa hold vast potential in adopting EVs. Sales of EVs in India doubled from 150,000 units in 2021 to 300,000 units in 2022, primarily on account of incentives from the government and affordable models. India will be one of the top three markets for EVs globally by 2030, and its annual sales will be over ten million units.

Table 9.3 shows the share of the Asia-Pacific region in the global market and the share of other regions in 2030. Europe and North America will

Table 9.3 Projected EV Sales by Region (2030).

Region	*Projected EV Sales (Millions)*	*Share of Global Market (%)*
Asia-Pacific	20.0	50
Europe	12.0	30
North America	6.0	15
Rest of the world	2.0	5

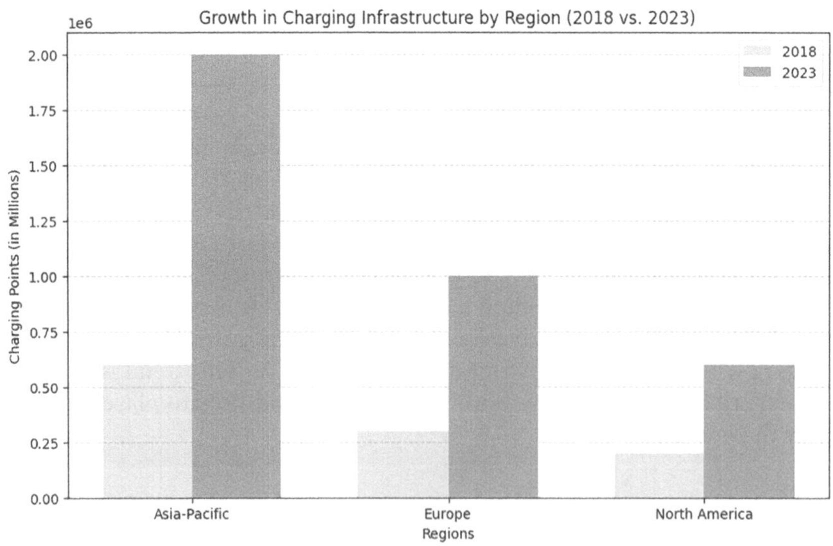

Figure 9.2 Growth in Charging Infrastructure Analysis Based on Regions.

continue to dominate the EV markets, while the rest of the world, including the emerging markets, will exhibit only moderate growth.

In conclusion, the global EV market is on a robust growth trajectory, driven by technological advancements, supportive policies, and increasing consumer awareness. While challenges remain, the market's evolution promises to redefine the automotive industry and contribute to a more sustainable future. The trends and projections discussed here underscore the transformative potential of EVs in reshaping global transportation.

9.3 GOVERNMENT INCENTIVES AND POLICIES

Government incentives and policies are among the most important factors moving the world forward toward EVs. Such measures constitute a foundation whereby the high front-end costs, the lack of infrastructures, and the consumer's uncertainty would be addressed. This section skims over the various initiatives from governments in all regions, the effects on the EV market, regional differences, and the direction courses.

Economic Incentives to the Purchaser of Electric Vehicles

Governments across the world have used several financial incentives to reduce the purchase price of EVs to make it reachable to customers. The financial incentives used have been direct subsidies, tax credits, and no fees on registration. The US government has dedicated as much as $7,500 in tax credit for a new EV in the United States, according to the Inflation Reduction Act of 2022. The rebates offered by states range between $2,000 and $5,000. California, New York, and Colorado offer the most rebate. These steps have considerably promoted the growth of EV sales in the United States, which increased to almost 800,000 units in 2022 from 330,000 units in 2018.

Financial incentives are led by Norway and Germany in Europe. Norway, where EVs made up 86% of new car sales in 2022, has exempted EVs from value-added tax (VAT) and registration fees and allowed them to use toll-free roads. Germany, the largest automotive market in Europe, offers subsidies of up to €9,000 for battery electric vehicles (BEVs) and €6,750 for plug-in hybrid electric vehicles (PHEVs), which has helped EVs reach 17% of total car sales in 2022.

Table 9.4 highlights the relationship between financial incentives and EV market penetration. Norway's generous subsidies have positioned it as a global leader in EV adoption, while Germany and China have also seen substantial growth due to their robust incentive structures. In contrast, emerging markets like India offer smaller subsidies, resulting in lower EV market shares.

Table 9.4 Financial Incentives for EVs in Selected Countries (2022).

Country	*Maximum Subsidy/Tax Credit (USD)*	*EV Share of New Car Sales (%)*
United States	$7,500	6.2
Norway	$10,000	86.0
Germany	$9,750	17.0
China	$6,000	25.0
India	$2,000	1.3

Table 9.5 Growth in Charging Infrastructure by Region (2018–2023).

Region	*Charging Points (2018)*	*Charging Points (2023)*	*Growth Rate (%)*
Asia-Pacific	600,000	2,000,000	233
Europe	300,000	1,000,000	233
North America	200,000	600,000	200

Policy Frameworks for Infrastructure Development

Financial incentives reduce the cost for the consumer to purchase, but an EV infrastructure is critical, particularly in charging networks, for mass adoption. Governments have used different strategies to promote investment in EV infrastructure, such as grants for charging stations, public–private partnerships, and urban planning initiatives.

China leads the world in developing EV infrastructure, with over 1.8 million public charging points as of 2023, bolstered by heavy investments from the government. The Chinese government has made it mandatory for 30% of new parking spaces in residential buildings to have charging stations. This was one of the key factors in achieving 25% EV market share in 2022.

The US NEVI plan has allocated $5 billion to support a nationwide public charging network within five years. The country has installed 50,000 public chargers by the end of 2023. Europe too is moving toward ultra-fast stations, and at present, more than 20,000 such stations were on ground in the Netherlands alone by the end of 2023.

Table 9.5 shows an explosion of all EV charging infrastructure within the Asia-Pacific and Europe regions. These regions had sound government policies and all funds in place for developing sets of infrastructures.

Regulatory Measures and Emission Standards

To hasten the shift from ICEVs to EVs, governments have taken strict regulatory measures, which include emission standards, mandating zero emission vehicles (ZEVs), and banning fossil fuels. The EU set a target of 95 grams of CO_2 per kilometer for new cars, so the EU encourages the auto

Table 9.6 Regulatory Measures Driving EV Adoption.

Country/Region	*Key Regulation*	*Target Year*	*Impact on Market Share*
European Union	ICE ban by 2035	2035	30% (2022) to 100%
United States	ZEV mandate in California and 15 states	2035	6% (2022) to 40% (2030)
China	NEV credit system mandating EV production	2025	25% (2022) to 40%
India	National Electric Mobility Mission Plan (NEMMP)	2030	1.3% (2022) to 30%

industry to develop EVs. Moreover, recently the EU committed to ban new sales of pure ICEVs by 2035; similar actions have already been undertaken by the United Kingdom and Canada.

In the United States, California has introduced a ZEV mandate, whereby, by 2035, all new passenger cars sold are expected to be ZEVs and deployments. This has been adopted by another 15 states, making a total of more than 40% of the US car market. Similarly, in China, NEVs are required through its NEV credit system [4–6] that sets a minimum requirement for annual numbers of EVs on sale to obtain credits.

Table 9.6 shows the aggressive goals set by governments to phase out ICEVs. While developed markets like the EU and the United States have put in place strict regulations, emerging markets like India are slowly aligning their policies to support EV growth.

Challenges in Policy Implementation

In spite of all incentives and policies implemented so far by the governments, much is to be done. EV adoption is hindered in most regions due to the lack of integrating policies with infrastructural inadequacy. For instance, take the Indian market for EVs. Its weakness is due to the poor disbursal of subsidies and poor charging infrastructure. In the United States, charging infrastructure remains poorly developed in regions outside major cities.

The challenge here is the balance of incentives with fiscal sustainability. High subsidies are becoming a burden on the government budget, as is already the case in Norway, where policymakers are considering cutting back EV incentives. Easy benefits should go to low-income families for wide-scale adoptions. The future of government incentives and policies for EVs lies in a balanced approach that combines financial incentives, infrastructure investment, and regulatory frameworks. With the decreasing battery price and continually advancing technology, governments can take away their subsidization and call for long-run initiatives like usage of renewable power and recycling of the battery.

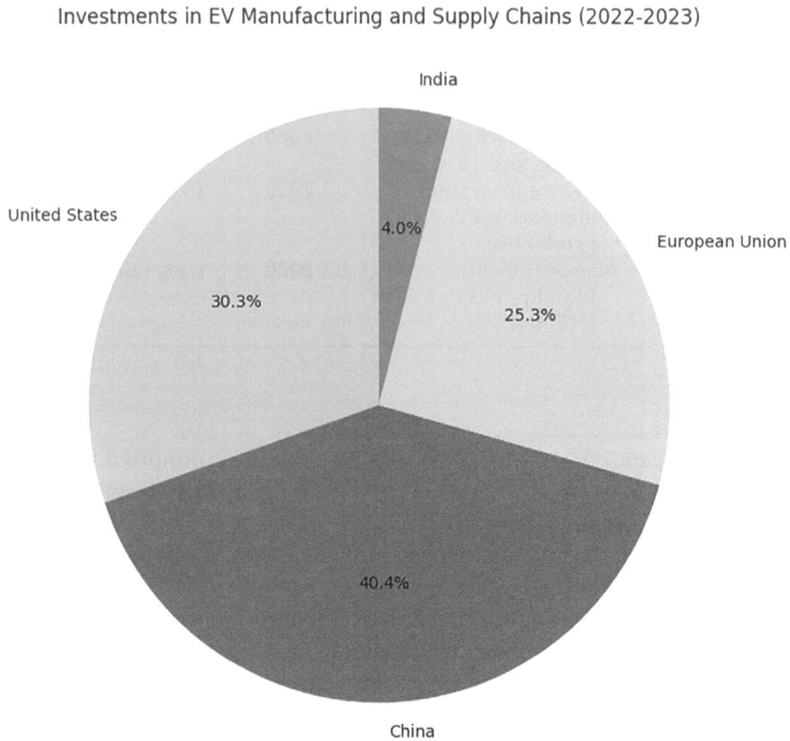

Figure 9.3 Investments in EV Analysis for Manufacturing Processes.

In recent history, government policies and incentives were critical for enabling speed in the direction of electric mobility. From pure financial incentives in the form of subsidies to comprehensive regulatory mandates across the board—and, lately, infrastructure-related mandates—these measures made all the difference along the growth curves of the burgeoning EV market. The data and some trends presented during this section again highlight the fundamental role of policies in overcoming bottlenecks.

Regulations That Encourage the Adoption of EVs

The regulation determines the course along which the EV market moves, and this course is inescapable. The regulatory policy adopted by various governments and international organizations, although aimed at achieving reductions in GHG emissions, supporting EV manufacturing, and pushing forward toward more eco-friendly modes of transport, would be looked into as an all-inclusive piece in the book, elaborating on the worldwide regulations that determine this market's industry impact on and the effects upon its growth going forward.

Emission Standards as Drivers of Change

One of the regulatory measures that motivate the adoption of EVs is through strict emission standards for vehicles. Such standards pressure the automakers to cut down on the emissions of their fleets, an incentive to produce and market EVs and hybrid vehicles. Since 2020, the CO_2 emission target for new cars in the EU has been set at 95 grams per kilometer with further cuts planned in the years 2030 and 2035. This caused massive investment in the development of EVs from the European automobile manufacturers [7–9] as few brands have announced that they would be stopping the sales of the pure ICEVs by 2035.

For instance, in the United States, Corporate Average Fuel Economy standards would mean that all automobile manufacturers must ensure a fleet average fuel efficiency of 49 miles per gallon in 2026; this is significantly higher than the current level of 36 miles per gallon. Such rules have become a driver for producing EVs and plug-in hybrids as companies are desperate to attain compliance standards to avoid penalty.

Zero-Emission Vehicle Mandates

ZEV mandates serve as another powerful regulatory push. California, acting as a bellwether before ZEV policy, has set the annual percentage for ZEV sales made by manufacturers. To date, 15 additional US states have passed the ZEV mandate, cumulatively accounting for more than 40% of US automobile sales. Under California's Advanced Clean Cars II rule, all new light-duty vehicles must be zero-emission in 2035.

China's NEV mandate is similar to the ZEV framework, where a set percentage of annual sales must be earned in the form of NEV credits. From 2023, the mandatory NEV credit ratio was still at 18%, and this ratio had gone up even higher to 20% by 2025 from recent observation. This policy really spearheaded China's EV production and sales quite remarkably, grabbing 25% of China's total vehicle sales in 2022.

Table 9.7 presents the aggressive goals of major automotive markets to switch to EVs. Regulatory norms from these auto markets will find their way to automaker plans and shape up technology for innovation in the world of EVs.

Bans and Phase-Outs for Fossil Fuel-Powered Vehicles

Increasingly, countries and cities are adopting policies to phase out fossil fuel-powered vehicles, thus speeding up the shift to EVs. Norway has been a leader in the world, having vowed to end the sale of new ICEVs by 2025. The EU recently announced it will end the sales of new ICEVs beginning in 2035. The United Kingdom and Canada also have made commitments to do likewise.

Table 9.7 Comparison of Emission and ZEV Mandates.

Country/Region	*Key Regulation*	*Target Year*	*Impact on EV Sales (%)*
European Union	95g CO_2 / km emission target	2030	17% (2022) to 50%
United States (CAFE)	49 MPG fuel efficiency target	2026	6% (2022) to 40%
California (ZEV)	100% ZEV sales	2035	15% (2022) to 100%
China (NEV)	20% NEV credit ratio requirement	2025	25% (2022) to 40%

Already, many cities have low-emission zones, which exclude older, higher-emission vehicles. These are being phased into zero-emission zones, in which only EVs and hydrogen-powered vehicles are allowed. Paris has announced it will restrict entry into its city limits to only ZEVs by 2030, a policy that already has investments in charging infrastructure and public EV fleets.

Incentives in EV Manufacturing and Supply Chains

Other than consumer-oriented regulations, most governments have implemented policies meant to encourage production of EVs and components. For instance, in the United States, the Inflation Reduction Act (IRA) of 2022 has contained incentives to stimulate automakers in building battery manufacturing facilities within their borders. This would reduce dependence on foreign supply chains and boost the production of similar items locally. US home production of battery cells or modules would then be eligible to accrue tax credits under the IRA, up to $45/kWh.

China has made some efforts by providing subsidies to the manufacturers of the batteries and to automakers for making EVs. As a result, China has emerged as the world's largest producer of EV batteries. In Europe, this process has already begun by adopting the Green Deal framework and initiating the European Battery Alliance for building up a competitive and sustainable battery industry in Europe. All of these are highly critical for the availability of EVs and to help in further supporting the global shift to electric mobility.

As per Table 9.8, most of the developed economies have made huge investments to drive the growth of EVs. The key bottlenecks that the investment in manufacturers and supply chains would address include battery shortages and high production costs.

Regulatory Hurdles

While regulations have been very good in terms of successes, several challenges persist. Emission standards and ZEV mandates involve very

Table 9.8 Investments in EV Manufacturing and Supply Chains (2022–2023).

Country/Region	*Investment in EV Manufacturing (USD Billions)*	*Key Initiative*
United States	15.0	Inflation Reduction Act
China	20.0	NEV Subsidy Program
European Union	12.5	European Battery Alliance
India	2.0	Production-Linked Incentive Scheme

strong monitoring and compliance mechanisms that are extremely resource-intensive. Among the low-income countries, strict institutional capacity and infrastructure constraints preclude the application of stringent regulations.

The shift to EVs has also been criticized for the unfair allocation [10–12] of benefits. For instance, various regions have recently come to appreciate that low-income families cannot afford EVs, which are priced high, and, more so, very few of the affordable options are available in the market. Thus, this demands an effort in rectifying disparities in the switch to sustainable mobility.

Future Regulation

In the future, regulatory frameworks in the EV market would change in terms of the new challenges and opportunities in the EV market. Integration of renewable energy sources in the EV market, development of vehicle-to-grid (V2G) systems, and recycling of EV batteries [13–15] are the core of the policy that the policymakers are trying to work out. Standardization of regions plays a major role in developing global trade and investment in the EV industry.

In a nutshell, it is regulatory pressure that has pushed car manufacturers to include EVs in their portfolios, along with changing consumer behavior. From emission standards to ZEV mandates and even fossil fuel bans, these steps formed a sound basis for shifting toward sustainable transport. This section's information and trends thus indicate that regulatory interventions will continue to be one of the significant measures to bypass obstacles and spur the global switchover to electric mobility.

REFERENCES

1. Dalkic-Melek, G., Saltik, E.C., & Tuydes-Yaman, H. Electric vehicle (EV) market penetration in countries with rising motorization rates. *Int J Civ Eng* (2024). https://doi.org/10.1007/s40999-024-01039-z

2. Qiu, Y., Deng, N., Wang, B. *et al.* Power supply disruptions deter electric vehicle adoption in cities in China. *Nat Commun* 15, 6041 (2024). https://doi.org/10.1038/s41467-024-50447-1
3. Agbesi, P.K., Ruffino, R., & Hakovirta, M. Creating an optimal electric vehicle ecosystem: An investigation of electric vehicle stakeholders and ecosystem trends in the US. *SN Bus Econ* 4, 28 (2024). https://doi.org/10.1007/s43546-024-00624-7
4. Javadnejad, F., Jahanbakh, M., Pinto, C.A. *et al.* Analyzing incentives and barriers to electric vehicle adoption in the United States. *Environ Syst Decis* 44, 575–606 (2024). https://doi.org/10.1007/s10669-023-09958-3
5. Kek, V., Kandasamy, J., Vezhavendhan, R. *et al.* Digital technology adoption model for electric vehicle battery recycling supply chain—an influential relationship mapping. *Oper Manag Res* 17, 1469–1508 (2024). https://doi.org/10.1007/s12063-024-00515-y
6. Singh, G., Misra, S.C., Daultani, Y. *et al.* Electric vehicle adoption and sustainability: Insights from the bibliometric analysis, cluster analysis, and morphology analysis. *Oper Manag Res* 17, 635–659 (2024). https://doi.org/10.1007/s12063-024-00442-y
7. Sonar, H., Belal, H.M., Foropon, C. *et al.* Examining the causal factors of the electric vehicle adoption: A pathway to tackle climate change in resource-constrained environment. *Ann Oper Res* (2023). https://doi.org/10.1007/s10479-023-05519-8
8. Pi, Z., & Wang, K. Operations and decarbonization of gasoline vehicle automakers under the dual credit policy: Green technology innovation and electric vehicle market entry strategies. *Ann Oper Res* (2023). https://doi.org/10.1007/s10479-023-05579-w
9. Ehsan, F., Habib, S., Gulzar, M.M. *et al.* Analyzing policy implications by considering adoption barriers on consumer adoption intention for electric vehicles: A comprehensive overview. *Environ Dev Sustain* (2024). https://doi.org/10.1007/s10668-024-05473-w
10. Mehmood, K., Qiu, X., Ghaffar, A. *et al.* Deciphering the future of electric vehicles amid emissions and adoption drivers. *Ambio* 53, 1686–1713 (2024). https://doi.org/10.1007/s13280-024-02026-3
11. Dhairiyasamy, R., Gabiriel, D., Bunpheng, W. *et al.* A comprehensive analysis of India's electric vehicle battery supply chain: Barriers and solutions. *Discov Sustain* 5, 361 (2024). https://doi.org/10.1007/s43621-024-00595-7
12. Lou, J., Shen, X., Niemeier, D.A. *et al.* Income and racial disparity in household publicly available electric vehicle infrastructure accessibility. *Nat Commun* 15, 5106 (2024). https://doi.org/10.1038/s41467-024-49481-w
13. Liang, J., Qiu, Y., Liu, P. *et al.* Effects of expanding electric vehicle charging stations in California on the housing market. *Nat Sustain* 6, 549–558 (2023). https://doi.org/10.1038/s41893-022-01058-5
14. Satrio, J., Juned, M., & Salam, S. International and domestic factors of battery electric vehicle technology diffusion in Japan. *East Asia* 41, 109–127 (2024). https://doi.org/10.1007/s12140-023-09418-4
15. Li, Y., Li, Q., & Shi, J. Effects of dual-credit policy on electric vehicle diffusion under subsidy transfer: A multi-stakeholder perspective. *Environ Dev Sustain* (2025). https://doi.org/10.1007/s10668-024-05926-2

Chapter 10

Future Developments

10.1 ADVANCEMENTS IN BATTERY TECHNOLOGY

Without question, evolution in the field of batteries has brought about an aggressive shift in electric vehicles (EVs). In the most drastic change, the world is moving its automobiles from internal combustion engine vehicles (ICEVs) to EVs, and batteries have been core in enabling this transition. It may be stated that the published future of electric mobility depends on the development of batteries that exceed in efficiency in energy densities, and have reduced charging times, lower costs, and more sustainable development. Research and industry are constantly exploring new materials, chemistries, or designs to improve performance, safety, and longevity to make EVs a redefined world. This is expected to facilitate the full-scale acceptance of EVs.

The most important aspect of battery advancement is energy density improvement. Energy density relates to how far an EV can travel, which is generally an important parameter concerning consumer adoption. Today, lithium-ion batteries constitute the leading market, with energy densities of approximately 250–300 Wh/kg. Efforts are underway to try and find ways of producing energy beyond that offered by current chemistries through research on emerging battery chemistries, including solid-state batteries and lithium–sulfur batteries. For example, if an EV equipped with a conventional lithium-ion battery pack of 60 kWh has a range of 400 km, a battery that boasts making it double the density energy could extend this capacity without increasing the weight of the assembly, 0 km, effectively eliminating range anxiety—the biggest concern associated with EVs.

Battery age and duration are other major improvements. Conventional lithium-ion batteries lose life through dendrites and instability of the electrolyte. On solid-state batteries, sometimes the replacement of liquid electrolyte with solid electrolytes has yielded better performance in terms of reduced degradation and longer lifespan. Regular lithium-ion batteries usually go through about 1,000–2,000 cycles before reaching 80% of their initial capacity. Solid-state batteries that can possibly surpass 5,000 cycles can replace them. Current technology allows the average lifespan of EV

DOI: 10.1201/9781003608097-10

batteries to be around 8–10 years; the lifespan of a solid-state battery could stretch well beyond 20 years, hence making EVs more economically viable in the long run.

Charging time is one dimension along which they make significant strides. Currently, fast charging networks permit charging to around 80% in 30–45 minutes, but ultrafast charging is being designed into the new battery chemistries. Lithium–iron phosphate (LFP) and silicon-anode batteries charge up to 80% in less than ten minutes. This is influenced by better improvements in how the electrodes are designed and the improved optimization of thermal management systems. This would create high potential savings on future charging infrastructure and increase consumer convenience by allowing EV recharging to match the speed of refueling a gasoline vehicle.

Turning used batteries into sustainable materials is yet another vital step forward. Greenhouse gases emitted by batteries and lifecycle aspects linked to mining lithium, cobalt, and nickel for batteries have profoundly troubled the conscience of society in regard to sustainability and the ethics of sourcing. All battery generations thus contend toward new battery solutions—be it sodium-ion or magnesium-ion batteries—guaranteed greater availability, and less environmental harm. Sodium-ion batteries exhibit quite similar electrochemical similarity as those of lithium-ion batteries while doing away with the need for expensive and rare materials. While sodium ion's density is slightly less, enhancements in cathode and anode design are likely to make them suitable for application in mass-market EVs.

Recycling and second-life applications for EV batteries are also undergoing significant development. Currently, less than 5% of lithium-ion batteries are recycled, thus contributing significantly to electronic waste and resource depletion. Advanced recycling techniques include hydrometallurgical and direct [1–3] recycling methods, techniques through which relevant materials can be recovered highly efficiently. Furthermore, used EV batteries with about 70% of potential usage are assigned for energy storage for renewable power grids. For example, a 40-kWh EV battery that is no more suited for car applications can still provide reliable energy storage for residential solar systems or commercial microgrids, extending its lifecycle by another decade.

Automotive instances alone do not represent the impact of these advancements; they cover a larger societal dimension. In many developing nations, reliable access to electricity is a challenge. Here, second-life EV batteries can feature transformative technologies for off-grid energy storage solutions. Communities without a stable grid connection in Africa and parts of Southeast Asia would use repurposed EV batteries to store solar or wind-generated energy that could assure a continuous supply of electricity. Finally, improved battery technologies reduce dependency on fossil fuels, translating into decreased carbon emissions and thereby toward assisting countries in achieving their climate goals under the Paris Agreement.

Affordability continues to be a considerable barrier to EV adoption, but continuous improvements in battery technology are steadily reducing the cost. Lithium-ion battery prices were around $1,000/kWh in 2010, and the price is now about $130/kWh. The prediction made by analysts regarding future costs does not vary much; scaling economies with advancements in new materials will see the cost drop below $60/kWh in a decade's time. To place this in context, a 75-kWh battery pack today costs $9,750; advancements in new materials would lower this to $4,500, allowing EVs to be available to the average consumer at a much better price.

Safety is also largely sought in terms of innovations in batteries. Currently, with thermal runaway known to cause fires in lithium-ion batteries, this underlying concern in raising serious safety concerns. Manufacturers have been adopting fire-resistant solid electrolytes, advanced cooling systems [4–6], and fail-safe designs that prevent internal short circuits in new advancements. Furthermore, new developments in battery management systems (BMSs) play an important role in monitoring battery health, predicting failures, and optimizing charging cycles to extend the lifespan while maintaining safety standards.

Adding artificial intelligence (AI) and machine learning battery technology is another groundbreaking development. With predictive analytics, AI can be used to analyze millions of real-time data for the optimization of energy usage, thermal management, and a better lifespan of batteries. Intelligent charging systems that ad-apt to user behavior and grid conditions are currently under development, and they would be charging electric vehicles at optimal times to reduce electricity costs and prevent potential grid overload.

Meanwhile, from a political point of view, the ongoing advancements in battery technologies are reshaping the emerging global economic and strategic theater. Countries heavily investing in battery research and production, for example, China, the United States, and Germany, are seen as future leaders in the newly developed fields of transportation and energy storage. This would lead their respective countries to reconsider heavily funding domestic supply chains, instead of depending on other countries' raw materials, and strengthening their energy security as foreign dependencies decrease.

New horizons in battery technology indeed are transforming the EV industry. Some critical challenges in the range of EVs include charging speed, cost, safety, and sustainability. Advances in high-energy-density chemistries, solid-state electrolytes, fast-charging solutions, sustainable materials, and AI-driven battery management would do their part in ensuring that the future would usher in vehicles that would be more efficient, accessible, and environmentally friendly. The maturation and scaling of such technologies would not only revolutionize the face of transportation but also have broader and deeper societal benefits, such as energy resilience, reduced carbon footprints, and the prospect of an efficient growing economy. The next decade will surely see unprecedented breakthroughs, making electric mobility the dominant paradigm in the automotive world.

10.2 AUTONOMOUS DRIVING AND ELECTRIC VEHICLES

Development is likely to create new linkages between autonomous driving and EVs; it promises revolutionary changes in efficiency, safety, and sustainability for the transportation industry. Hand in hand with the leaps made by electric mobility comes autonomous driving technology, driven by AI combined with sensor integration and machine learning. Co-dependence of these two technological spheres will mean a redesign of urban transportation, logistics, and personal mobility. One example is that self-driving systems will renew their capabilities, regarding energy consumption, traffic efficiency, and environmental sustainability, through their inclusion in EVs.

Autonomous driving is one of the basic reasons why the makings of the electric powertrain are highly compatible with the independent control systems that it has. For all types of vehicles, moving all their internal combustion engines (ICEs) involves complex mechanical architectures with many moving parts compared to an EV. Thus, it is generally accepted to have a higher degree of fine control over acceleration, braking, and energy management, which, in turn, improves the performance of self-driving algorithms. Since self-driving uses real-time data processing and decision-making to navigate the roads safely, features exclusive to EVs have direct energy optimization impacts. For example, self-driving algorithms are estimated to save an additional 20% of EV energy consumption, thereby extending range and reducing total energy consumption.

Take, for example, a typical EV with 75 kWh of battery energy content and an efficiency of 5 miles/kWh: if a human driver is in charge, this is likely to drop to about 4 miles/kWh because of all the acceleration and deceleration associated with human driving, yielding an overall range of 300 miles. An autonomous system optimizing acceleration, maintaining constant speed, and eliminating sudden braking could boost this up to 5.5 miles/kWh, raising the total range to about 412 miles. Such changes dramatically lessen the range anxiety—the primary concern for ongoing EV users—and will make long-distance EV travel even more compelling.

The economic implications of introducing autonomy in EVs are considerable. Fleet operators, particularly ride-sharing services and logistics companies, will significantly benefit from self-driving EVs. Without human drivers on their books, fleet operational costs for companies like Uber and Lyft are expected to drop by almost 60%. The cost incurred by autonomous electrified ride-sharing consumers is further estimated to significantly reduce per mile. Table 10.1 details the cost feature of traditional human-driven taxis, human-driven EV ride-sharing services, and totally autonomous EV ride-sharing services.

This table indicates that fully autonomous EVs can drop the cost per mile to one-sixth of that associated with gasoline taxis. This makes ride-sharing quite affordable, enhancing the prospects of large-scale acceptance and reducing private ownership of automobiles. Traffic congestion, parking demand, and emissions are thus expected to taper off in urban centers.

Table 10.1 Cost Per Mile of Different Ride-Hailing Models.

Ride-Hailing Model	*Human Driver Salary (per mile)*	*Energy Cost (per mile)*	*Maintenance Cost (per mile)*	*Total Cost (per mile)*
Human-driven gasoline taxi	\$0.70	\$0.12	\$0.15	\$0.97
Human-driven EV	\$0.70	\$0.06	\$0.10	\$0.86
Autonomous EV	\$0.00	\$0.06	\$0.08	\$0.14

Table 10.2 Energy Efficiency of Public Transportation Modes.

Transportation Mode	*Energy Consumption (kWh per passenger mile)*	*CO_2 Emissions (g per passenger mile)*	*Average Passenger Capacity*
Gasoline bus	0.50	250	40
Electric bus	0.30	50	40
Autonomous electric shuttle	0.22	10	10

Autonomous EVs prove to be very useful when it comes to safety; this is yet another important driving factor behind their acceptance. Human error accounts for about 94% of all traffic accidents in the world. Autonomous vehicles use advanced sensors such as LiDAR, radar, and cameras to recognize and respond to potential hazards much more accurately than human drivers. Self-driving systems eliminate risks related to distractions from the driver, fatigue, and intoxication, thereby providing a higher chance of reducing accidents. The introduction of Level 4 and Level 5 autonomous cars—where a car drives without any human input—could lead to a massive drop in 90% of all traffic deaths. If introduced on a global level, this would mean saving a rough estimate of 1.2 million lives every year.

Besides safety, the other considerations surrounding autonomous EVs relate to the potential losses of employment and subsequent displacement of the workforce. Millions of jobs in driving-related industries, including trucking, taxis, and delivery services, may be at stake as automation replaces human labor. Autonomously operated freight transport may impact the livelihoods of more than 3.5 million truck drivers in the United States alone. Table 10.2 compares the energy efficiency of various modes of public transport while illustrating the operational advantages of autonomous EVs.

This table illustrates that autonomous electric shuttles consume less energy per passenger mile and produce significantly lower carbon emissions compared to traditional buses. As cities continue investing in sustainable transportation solutions, self-driving EVs will most likely become an integral part of urban mobility ecosystems.

The impact of autonomous EVs on the environment extends far beyond energy efficiency and emission reductions. Their very widespread adoption would create the conditions for the fleet of self-driving EVs to effectively integrate renewable energy sources into their transport demand. For example, autonomous EVs are designed or programmed to charge at times when the renewable energy input is at its maximum, such as solar and wind. This approach could mean less reliance on fossil fuel electricity generation and creating the groundwork for a sustainable energy grid. The vehicle-to-grid (V2G) technology would enable autonomous EVs to act as mobile energy storage units by injecting excess electricity into the grid during demand peaks.

The societal implications of autonomous EVs are likewise quite far-reaching. The disabled, elderly, and those without driver's licenses would all highly benefit from the added accessibility and enhanced mobility options offered to them. Autonomous EVs can provide on-demand transport services customized for users with limited mobility, thus promoting independent living and fostering social inclusion. In some cases, in areas where such transport is pale, self-driving EVs could help connect underprivileged regions and advance opportunities.

Despite this amenable scenario, as concerns remain to be addressed for mass deployment of autonomous EVs, multiple hurdles are yet to be traversed. It remains to lay down proper rules that ensure the safety, address the liability, and guide the ethical development of this self-driving technological wonder. Another concern is cybersecurity, considering autonomous vehicles come with complex networks of sensors, communication systems, and cloud-based algorithms, all of which might be hacked. Almost equally important is the yet-dichotomous public perception of autonomous driving, which varies in trust for AI-driven mobility along demographic and regional lines.

To summarize, the endless adventures and odysseys in the field of autonomous vehicles mixed with EVs truly mark an ontological shift in the mobility industry. This beautiful amalgamation of clean energy propulsion with intelligent automation would offer mammoth advantages along with better cost, energy efficiency, safety, and ecological integrity. With further technological development, autonomous EVs will remain the fulcrums for the reimagination of mobility across urbanity, public transport, and human-powered transport in the entire food chain. While there are challenges yet to be tackled, over the long term, the savings outweigh the entire possible impediments, and thus autonomous electric mobility will be an undeniable panacea in the years ahead.

10.3 VEHICLE-TO-GRID INTEGRATION

The V2G technology signifies a fundamental change in the energy and transportation sectors as a means of establishing a bidirectional power flow between EVs and the power grid. While the use of EVs increases throughout the world, the adoption of the V2G systems will benefit the general stability of the grid in terms of renewable energy integration and value paid to owners for vehicles. By using energy stored in EV batteries [7–9], V2G technology can bring balancing purposes in demarcation lines between demand and supply of electricity, relieving peak hour stress for the grid, and allowing for a much wider exploitation of renewables within energy sourcing processes.

The point of convergence in energy flows is essential in V2G integration: EVs have not only the attribute of consuming electricity but also that of acting as mobile energy storage. When there is low electricity consumption in the system, EVs collect an overabundance of renewable power, such as solar and wind energy. Whenever peak load arises, EVs can return energy to the grid, thus helping keep voltage and frequency stable. Therefore, this bidirectional power flow makes EVs an essential part of smart grid infrastructure, which will limit dependency on fossil fuel peaker plants typically used to manage high demands.

Enabling V2G will mean earning some cash for most in participating programs, but this involves evaluating the revenue potential to make sure how economically viable it is going to be for the V2G technology user. Depending on factors like battery size, grid interaction frequency, and electricity rates, the EV owner can generate income through stored energy resale on the grid. Table 10.3 is an illustration of potential annual earnings for an EV owner based on grid participation levels.

If the EV owner empties 5 kWh/day at $0.10/kWh, that is a little over $182.50 earned every year. The maximum discharge amount, at 15 kWh/day and $0.15/kWh, nets more than $800 in profit per year. This kind of economic incentive might prompt a considerable number of EV owners to involve themselves in these programs and make the ownership of EVs much more attractive in financial terms.

One other important benefit that V2G technology might have is integration with renewables. The big problem with solar and wind energy is that one cannot rely on these resources all the time; the amount of electricity produced is dependent upon the occurrence of the weather, leading to renewable generation uncertainty and the credibility of the grid. At periods of too high generation, even during production decline periods, discharging energy stored in EVs enabled with V2G could supply the power grid with excess renewable energy. For example, the maximum generation of

Table 10.3 Potential Annual Earnings From V2G Participation.

V2G Participation Level	*Average Daily Discharge (kWh)*	*Electricity Selling Price ($/kWh)*	*Annual Earnings ($)*
Low	5	0.10	182.50
Medium	10	0.12	438.00
High	15	0.15	821.25

electricity from solar power is at noon hours when the electricity demand is low. That is when EVs can recharge and return energy into the grid during evening hours when the demand peaks.

The impact of V2G and the stability of the grid at the peak level are of greatest relevance in regions that have high penetration of renewable energy sources. Denmark, the Netherlands, and Germany are the forerunners of the world in the use of V2G technology in national energy markets where the integration of EVs into the market economy is taking place. In Denmark, for example, fleet operators now have access to V2G charging stations where they can provide grid services, eliminating the need for extra energy storage infrastructures. Arguably, this strategy could add more value to other areas with a higher concentration of EVs, optimizing energy distribution and reducing transmission losses.

To further show the importance of V2G in improving grid reliability, the following Table 10.4 compares stability parameters with and without V2G integration. Approximately 20% of the present peak load could become possible with V2G implementation, relieving strain on the power grid. Frequency deviations causing spark instabilities would improve from approximately ±0.20 Hz to ±0.05 Hz or 75% levels. This improved regulation of frequency further curtails outages and voltage instability. Also, increased V2G technology causes the utilization of renewable energy to increase by 27%, creating an ecosystem more sustainable for the grid and less dependent on fossil fuel plants.

The advancement of society through technical means is significant, as V2G will have equally strong social and environmental effects; in reducing dependence on fossil fuel energy generation, V2G also contributes directly to climate change mitigation by taking away carbon emissions. If V2G is widely adopted by enough households to lower peak demand by 10%, that would mean a large amount of coal- and gas-fired power plants would not be necessary, leading to an annual CO_2 emission reduction of several million metric tons. That indeed fits into the international climate targets that are now included in the Paris Agreement, where nations have committed to greenhouse gas emission reductions.

Under their social aspect, V2G technology brings about increased access to unequally distributed energy and decentralized energy distribution. For

Table 10.4 Impact of V2G Integration on Grid Stability.

Grid Stability Metric	*Without V2G*	*With V2G*	*Improvement (%)*
Peak load reduction (MW)	2,000	1,600	20%
Frequency fluctuation (Hz)	±0.20	±0.05	75%
Renewable energy utilization (%)	55	70	27%
Blackout risk reduction (%)	0	10	-

areas where electricity is not a dependable commodity in the residential areas, EV owners can use their vehicles as a substitute power source, providing electricity to homes during power failures. This is particularly beneficial where there are possibilities of natural disasters disrupting power grids. For example, in the aftermath of hurricanes or earthquakes, V2G-enabled EVs could be deployed as emergency power units, ensuring that critical infrastructure such as hospitals and communication networks remains operational.

In addition, V2G can assist low-income communities in its cost-reduction programs. V2G enables households to store and sell their excess electricity, promoting social equity in energy and reducing centralized dependability on utility providers. Community V2G programs operating under the premise that neighborhoods collectively harness energy storage and distribution may lower electricity prices while providing additional monetary incentives to EV purchase in disadvantaged regions.

With a bright future ahead, challenges to bulk applicability must be addressed. Battery degradation tops this list: with frequent charging and discharging cycles, the lifespan of batteries may reduce. Nevertheless, with advances in battery technology, especially solid-state batteries, this concern may be resolved with a longer cycle life and improved thermal stability. These in turn call for establishing regulatory and market frameworks in such a way that V2G participation would be standardized for equitable compensation to EV owners, thus preempting any ill-intent by utility providers.

Another obstacle is in widespread deployment of the infrastructure. V2G-compatible charging stations need sophisticated bidirectional inverters, smart grid communication systems, and sophisticated software algorithms

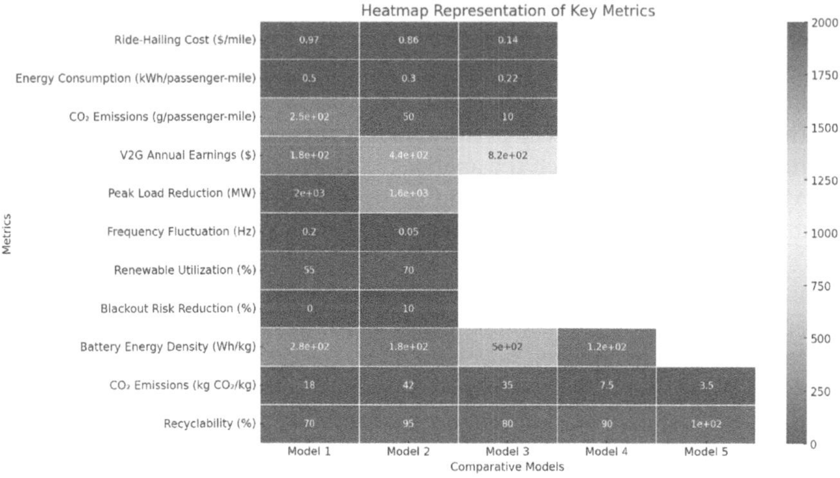

Figure 10.1 Battery Parameter Conclusive Analysis.

to efficiently manage energy flow. The initial costs of such infrastructures might be high, but as technology develops and user adoption increases, prices should come down through economies of scale. In any case, public–private partnerships will greatly aid in fast-tracking V2G infrastructural development, where governments can encourage investment in smart grids and bidirectional charging networks.

To sum up, V2G systems can greatly revolutionize both energy and transportation realms. Managing bidirectional energy flow between EVs and the grid enhances system stability, provides economic benefits for EV owners, and maximizes renewable energy utilization. The data generated in this discourse shows one substantial effect of V2G in energy markets along reducing peak load, frequency control enhancement, and overall increased penetration of renewable energy. On the social dimension, V2G may enhance energy access, catalyze disaster resilience recovery, and de facto ultimately lessen electricity costs for communities that need it the most. There are other issues, including battery degradation, regulatory framework, and infrastructural needs, but all of these are in the way of V2G reaching a wider foothold in the coming future with the innovative battery technology and grid modernization. While countries worldwide are heading toward sustainable energy systems, V2G has a pivotal role to play, alongside electric mobility, in obtaining a clean, resilient, and financially viable future.

10.4 MATERIALS AND INNOVATIONS

The development of EVs has proceeded in parallel with the evolution of materials and technologies. As global demand ramps up for EVs, manufacturers and researchers seek to develop improvements in vehicle performance, energy efficiency, and sustainability through new materials and engineering breakthroughs. Materials in EV technology span beyond battery chemistry to include the structural body, lightweight composites, thermal management systems, and electronic components. The spin of materials directly ties in with vehicle weight, battery performance, safety, and recyclability, making material science very much the anchor to the evolution for EVs.

Most critical material innovations for EVs are in battery technology. Lithium-ion batteries, the energy storage system of choice today for most EVs, depend on lithium, nickel, cobalt, and graphite. Anyway, material scarcity, exposure to cost, and environmental damage are some of the factors propelling the research into all kinds of alternative battery chemistries. LFP batteries appear to fare well since they do not incorporate cobalt in their battery structure, and they have gained traction simply because they are cheap and able to maintain an excellent thermal stability. Similarly, solid-state batteries gain traction since they replace liquid electrolytes with solid materials, which promise higher energy density, improved safety, and lifespan. When solid-state batteries achieve energy densities over 500 Wh/kg against the standard lithium-ion battery, the latter will barely cross a range of 250–300 Wh/kg. An increment in the energy density simply implies longer distances covered, thus alleviating range anxiety from the mind of consumers.

To illustrate the relationship between various battery chemistries and EV performance, Table 10.5 outlines several comparisons between some crucial battery materials.

From the data shown, solid-state batteries would indisputably excel in energy density and cycle life over lithium-ion batteries and LFP batteries. However, their higher cost still poses a barrier to their extensive application. In this context, sodium-ion batteries are a reasonable option and potentially inexpensive since the raw materials are abundant and cheaper compared to lithium, especially for entry-level EVs and grid energy storage applications.

Apart from batteries, what is involved in EV innovation is largely structural material. Vehicle weight reduction is a great way to increase energy efficiency and maximize range. Steel is the main structural material for traditional combustion engine vehicles, giving them a heavy weight. And now, more and more, EV manufacturers are starting to say they are looking into lightweight materials such as aluminum, carbon fiber composites, and high-strength plastics to enhance performance without giving up on safety. A 10% reduction in weight might actually improve energy efficiency by as much as 6–8%, increasing driving range on a single charge.

Tesla strikes fair balance between weight and durability by employing hybrid aluminum and high-strength steel in its vehicle body structure. Given its aluminum-intensive bodywork, Tesla Model S weighs less than most other luxury sedans without sacrificing strength, while other manufacturers like BMW have employed carbon-fiber-reinforced polymer (CFRP) materials in their electric models (like BMW i3) to provide lightweighting. Modern materials, however, are few; they tend to be highly expensive and complexly fabricated. For example, carbon fiber composites deliver strength-to-weight ratios that are very good but cost much more than aluminum or steel, making them unsuitable for mass-market EVs.

Table 10.5 Comparison of Battery Chemistries in EVs.

Battery Type	*Energy Density (Wh/kg)*	*Cycle Life (Charge–Discharge Cycles)*	*Cost per kWh ($)*	*Safety Rating*
Lithium ion (lithium nickel manganese cobalt oxide)	250–300	1,000–2,000	130–150	Moderate
LFP	160–200	2,500–4,000	90–120	High
Solid-state	500+	5,000+	200–300	Very high
Sodium ion	100–150	2,000–3,000	60–80	Moderate

Innovation continues on materials and applies to thermal management systems too. Batteries in EVs generate heat during charging and discharging, and effective thermal control maintenance is essential for performance and safety. Traditional cooling systems are based on liquid cooling and new materials like phase change materials (PCM) and nanomaterials. PCMs provide passive cooling because they absorb and release thermal energy during phase transitions without needing any additional work. Studies are still undergoing investigation to learn how graphene heat spreaders improve thermal conductivity since overheating causes most damage to a battery.

One of the emerging areas of material innovation is electric motor performance enhancement. Rare-earth elements such as neodymium and dysprosium are the ones upon which the commonly used permanent magnet motor in the EV relies. Such alternatives in motor design have led to this diversification of motor topology, owing to disruption in supply and also due to geopolitical issues. Some models of Tesla, for example, use induction motors of higher efficiency, which do not utilize rare-earth materials. Studies into alternative magnetic materials, specifically iron-nitride-based compounds, are currently being undertaken, hoping to lessen the burden of rare-earth dependence while boosting motor performance.

Table 10.6 shows that social implications of material innovation in EVs concern performance within the vehicle, but more than that, they concern sustainability. Mining and extraction of such key materials as lithium, cobalt, and nickel raise ethical and environmental concerns. Most derived from the Democratic Republic of Congo, mining has traditionally met criticism for human rights abuses and environmental detriment. Therefore, EV manufacturers are actively looking for alternatives for ethical sources and invest heavily in recycling of batteries in order to be able to recover expensive materials from the used batteries.

Accordingly, data indicated in Table 10.7 that cobalt takes the first spot regarding CO_2 emissions and ethical issues, while sodium-based materials offer a much more sustainable and recyclable handling process. Very easily recyclable materials such as graphite and cobalt accentuate the need for an efficient recycling system to limit damage to the environment.

Table 10.6 Environmental Impact of EV Battery Materials.

Material	*CO_2 Emissions per kg (kg CO_2)*	*Recyclability (%)*	*Ethical Concerns*
Lithium	15–20	60–80	Moderate
Cobalt	35–50	95	High
Nickel	30–40	80	Moderate
Graphite	5–10	90	Low
Sodium	2–5	100	Low

Table 10.7 Review of Existing Models.

Reference	*Method*	*Main Objectives*	*Contextual Findings*	*Limitations*
[10] Bauer et al. (2015)	Lifecycle assessment	Evaluate environmental performance of passenger vehicles	EVs have lower lifecycle emissions than ICEVs	Scenario-dependent assumptions
[11] Deng & Lee (2019)	Graphene-based materials study	Investigate graphene use in energy storage	Graphene improves battery energy density and cycle life	High production cost
[12] Fang & Peng (2015)	Lithium–sulfur battery analysis	Evaluate rechargeable lithium–sulfur batteries	Higher energy density than lithium ion but with shorter cycle life	Electrolyte stability issues
[1] Gao & Zhou (2018)	High-voltage lithium-ion batteries	Assess advancements in high-voltage lithium-ion batteries	Increased voltage improves energy capacity	Safety and thermal stability concerns
[2] Hannan et al. (2017)	Review of energy storage	Examine challenges in EV battery storage	Thermal management is critical for battery longevity	Limited review of alternative chemistries
[3] Jiao & Bruce (2016)	Electrode material study	Analyze MnO_2 as a rechargeable lithium battery electrode	Mesoporous MnO_2 enhances charge retention	Limited application in commercial batteries
[4] Kang & Ceder (2009)	Ultra-fast charging–discharging study	Examine materials for rapid charge–discharge	Certain lithium-ion materials enable ultra-fast charging	Potential degradation over time
[5] Li et al. (2018)	Review of lithium-ion battery history	Summarize 30 years of lithium-ion development	Continuous improvements in energy density and safety	Limited discussion on alternative technologies
[6] Liu & Zhang (2015)	Nanostructured electrode materials	Analyze oxides and sulfides for lithium-ion batteries	Nanostructured electrodes enhance performance	Manufacturing complexity

(Continued)

Table 10.7 (Continued)

Reference	*Method*	*Main Objectives*	*Contextual Findings*	*Limitations*
[7] Manthiram (2020)	Lithium-ion cathode chemistry	Examine trends in cathode materials	High-nickel cathodes improve capacity	Nickel sourcing and stability issues
[8] Nitta et al. (2015)	Lithium-ion battery material analysis	Assess current and future lithium-ion materials	Graphene and silicon show promise	Limited large-scale applications
[9] Peters et al. (2017)	Environmental impact of batteries	Analyze lifecycle emissions of lithium-ion batteries	Recycling can significantly reduce emissions	Recycling infrastructure is limited
[13] Scrosati & Garche (2010)	Lithium battery status review	Discuss future prospects of lithium batteries	Solid-state and lithium--sulfur are promising	Commercial viability not yet achieved
[14] Tarascon & Armand (2001)	Lithium battery challenges	Identify research challenges in lithium batteries	Cathode stability is a major concern	Cycle life limitations
[15] Thackeray et al. (2012)	Beyond lithium-ion batteries	Explore post-lithium-ion technologies	Sodium ion and magnesium ion are viable alternatives	Lower energy density than lithium ion
[16] Wang & Sun (2015)	Carbon coating on $LiFePO_4$	Study the effect of carbon coatings on battery life	Coatings improve conductivity and longevity	Scaling production remains a challenge
[17] Xu (2014)	Electrolyte research	Analyze electrolytes for lithium-ion batteries	Stable electrolytes extend battery life	Toxicity and production cost concerns
[18] Yang et al. (2011)	Green grid energy storage	Investigate energy storage for renewable grids	EVs can support grid stability via V2G	Regulatory and infrastructure barriers
[19] Yoshino (2012)	History of lithium-ion battery	Document invention and progress of lithium-ion batteries	Fundamental principles remain unchanged	Limited scope beyond lithium ion
[20] Cano et al. (2018)	Battery and fuel cell review	Compare batteries and fuel cells for EVs	Fuel cells provide longer range	High hydrogen infrastructure cost
[21] Choi & Aurbach (2016)	Post- lithium-ion battery analysis	Evaluate high-energy-density alternatives	Lithium–air batteries have high potential	Poor cycle life and rechargeability

[22] Dunn et al. (2015)	Battery lifecycle study	Analyze energy and emissions of lithium-ion batteries	Recycling reduces overall impact	High recycling cost
[23] Fan et al. (2020)	Recycling technology review	Discuss challenges in lithium-ion battery recycling	Chemical recovery methods improve efficiency	Limited commercial implementation
[24] Gao & Zhou (2019)	Organic electrode materials	Explore organic-based battery electrodes	Potentially more sustainable than metal-based	Lower conductivity
[25] Hao et al. (2017)	GHG emissions from lithium-ion batteries	Quantify emissions in Chinese lithium-ion battery production	Production emissions are substantial	Regional variation in energy sources
[26] Huang et al. (2020)	Materials for lithium-ion batteries	Review advanced lithium-ion materials	Graphene and nanomaterials improve performance	Manufacturing scalability is a challenge
[27] Jian & Ji (2019)	Sodium-ion battery review	Assess cost-effectiveness of sodium-ion batteries	Lower cost than lithium ion but lower energy density	Limited commercial adoption
[28] Kim et al. (2014)	Aqueous rechargeable batteries	Investigate water-based lithium- and sodium-ion batteries	Safer and cheaper alternatives	Lower energy density
[29] Li et al. (2020)	High-nickel cathode study	Examine nickel-rich cathodes for EVs	Higher energy density but stability issues	Nickel supply constraints
[30] Liu et al. (2010)	Advanced materials for energy storage	Review future battery material developments	Nanostructures improve electrode efficiency	Scalability remains uncertain

Material innovations in EVs directly affect how affordable and accessible electric mobility becomes. Using cheaper materials with new manufacturing techniques may reduce the costs incurred in producing EVs and consequently make it possible for many consumers to afford them. Low-cost EVs could be produced for emerging markets, thanks to sodium-ion batteries that will cut battery costs drastically. 3D printing technologies and advanced composite materials are, therefore, poised to reduce material wastage during manufacture and lower production costs while allowing for manufacturability of complex yet lightweight architectures.

The charging infrastructure has to go along the same line with the materials. The advent of EVs must have a charging infrastructure compatible with the upcoming battery chemistries. Ultra-high charging networks will interface with an emerging super-refueling regime, like that available for ICEVs but instead relies upon high-temperature-resistant materials and advanced conductors. Emerging charging cable technologies based on conductive polymers and superconducting materials are being considered for future generations in an effort to minimize energy losses and improve charging applications.

These material improvements can best be interpreted within the premise of good government policy and sustainable industrial cooperation. Most governments around the world are motivating the advancement of research and development for renewable materials through grants, subsidies, and regulations. For instance, the EU has now fixed a compulsory target for recycling of batteries from EVs, where manufacturers will recover a certain share of the battery materials at the end of life. The US Department of Energy is also contributing toward the setting up of a domestic supply chain for critical materials to avoid dependence on foreign sources.

In a nutshell, materials and innovations shaping the future world of sustainable transport in EVs are battery chemistry, lightweight materials, thermal management, and motor technology. Improvements in these areas directly increase the performance of the vehicle and meet environmental and ethical concerns presented in the data. Material selection is, indeed, critical for energy efficiency, recyclability, and cost-effectiveness. Material innovations will socially make it economically accessible, reduce mineral dependence from conflict, and strengthen energy sustainability. While there are still some issues, research and collaboration continue to promote progress toward creating a more efficient, affordable, and green EV ecosystem. Thus, it will propel the clean future forward, bringing the next generation of EVs closer to being sustainable, high-performing, and accessible.

REFERENCES

1. Gao, H., & Zhou, H. (2018). Progress in high voltage lithium Ion batteries. *Energy & Environmental Science*, *11*(4), 1012–1027. https://doi.org/10.1039/C7EE03608K

2. Hannan, M. A., Hoque, M. M., Mohamed, A., & Ayob, A. (2017). Review of energy storage systems for electric vehicle applications: Issues and challenges. *Renewable and Sustainable Energy Reviews*, *69*, 771–789. https://doi.org/10.1016/j.rser.2016.11.171
3. Jiao, F., & Bruce, P. G. (2016). Mesoporous crystalline β-MnO2—a reversible positive electrode for rechargeable lithium batteries. *Advanced Materials*, *19*(5), 657–660. https://doi.org/10.1002/adma.200600877
4. Kang, B., & Ceder, G. (2009). Battery materials for ultrafast charging and discharging. *Nature*, *458*(7235), 190–193. https://doi.org/10.1038/nature07853
5. Li, M., Lu, J., Chen, Z., & Amine, K. (2018). 30 years of lithium ion batteries. *Advanced Materials*, *30*(33), 1800561. https://doi.org/10.1002/adma.201800561
6. Liu, J., & Zhang, J.-G. (2015). Nanostructured metal oxides and sulfides for lithium ion batteries. *Journal of Power Sources*, *282*, 492–508. https://doi.org/10.1016/j.jpowsour.2015.02.054
7. Manthiram, A. (2020). A reflection on lithium ion battery cathode chemistry. *Nature Communications*, *11*(1), 1550. https://doi.org/10.1038/s41467-020-15355-0
8. Nitta, N., Wu, F., Lee, J. T., & Yushin, G. (2015). Li ion battery materials: Present and future. *Materials Today*, *18*(5), 252–264. https://doi.org/10.1016/j.mattod.2014.10.040
9. Peters, J. F., Baumann, M., Zimmermann, B., Braun, J., & Weil, M. (2017). The environmental impact of Li ion batteries and the role of key parameters—A review. *Renewable and Sustainable Energy Reviews*, *67*, 491–506. https://doi.org/10.1016/j.rser.2016.08.039
10. Bauer, C., Hofer, J., Althaus, H.-J., Del Duce, A., & Simons, A. (2015). The environmental performance of current and future passenger vehicles: Life cycle assessment based on a novel scenario analysis framework. *Applied Energy*, *157*, 871–883. https://doi.org/10.1016/j.apenergy.2015.01.019
11. Deng, D., & Lee, J. Y. (2019). Graphene-based materials for energy storage and conversion. *Journal of Materials Chemistry A*, *7*(21), 12623–12650. https://doi.org/10.1039/C9TA02906A
12. Fang, X., & Peng, H. (2015). A revolution in electrodes: Recent progress in rechargeable lithium-sulfur batteries. *Small*, *11*(12), 1488–1511. https://doi.org/10.1002/smll.201402680
13. Scrosati, B., & Garche, J. (2010). Lithium batteries: Status, prospects and future. *Journal of Power Sources*, *195*(9), 2419–2430. https://doi.org/10.1016/j.jpowsour.2009.11.048
14. Tarascon, J.-M., & Armand, M. (2001). Issues and challenges facing rechargeable lithium batteries. *Nature*, *414*(6861), 359–367. https://doi.org/10.1038/35104644
15. Thackeray, M. M., Wolverton, C., & Isaacs, E. D. (2012). Electrical energy storage for transportation—approaching the limits of, and going beyond, lithium Ion batteries. *Energy & Environmental Science*, *5*(7), 7854–7863. https://doi.org/10.1039/C2EE21892E
16. Wang, J., & Sun, X. (2015). Understanding and recent development of carbon coating on LiFePO4 cathode materials for lithium ion batteries. *Energy & Environmental Science*, *8*(4), 1110–1138. https://doi.org/10.1039/C4EE04041D
17. Xu, K. (2014). Electrolytes and interphases in Li ion batteries and beyond. *Chemical Reviews*, *114*(23), 11503–11618. https://doi.org/10.1021/cr500003w

18. Yang, Z., Zhang, J., Kintner-Meyer, M. C. W., Lu, X., Choi, D., Lemmon, J. P., & Liu, J. (2011). Electrochemical energy storage for green grid. *Chemical Reviews*, *111*(5), 3577–3613. https://doi.org/10.1021/cr100290v
19. Yoshino, A. (2012). The birth of the lithium ion battery. *Angewandte Chemie International Edition*, *51*(24), 5798–5800. https://doi.org/10.1002/anie.201105006
20. Cano, Z. P., Banham, D., Ye, S., Hintennach, A., Lu, J., Fowler, M., & Chen, Z. (2018). Batteries and fuel cells for emerging electric vehicle markets. *Nature Energy*, *3*(4), 279–289. https://doi.org/10.1038/s41560-018-0108-1
21. Choi, W., & Aurbach, D. (2016). Promise and reality of post-lithium ion batteries with high energy densities. *Nature Reviews Materials*, *1*(4), 16013. https://doi.org/10.1038/natrevmats.2016.13
22. Dunn, J. B., Gaines, L., Kelly, J. C., James, C., & Gallagher, K. G. (2015). The significance of Li ion batteries in electric vehicle life-cycle energy and emissions and recycling's role in its reduction. *Energy & Environmental Science*, *8*(1), 158–168. https://doi.org/10.1039/C4EE03029J
23. Fan, E., Li, L., Wang, Z., Lin, J., Huang, Y., Yao, Y., & Wu, F. (2020). Sustainable recycling technology for Li ion batteries and beyond: Challenges and future prospects. *Chemical Reviews*, *120*(14), 7020–7063. https://doi.org/10.1021/acs.chemrev.9b00535
24. Gao, X., & Zhou, H. (2019). Towards sustainable and versatile energy storage devices: An overview of organic electrode materials. *Energy & Environmental Science*, *12*(3), 841–864. https://doi.org/10.1039/C8EE03679A
25. Hao, H., Mu, Z., Jiang, S., Liu, Z., & Zhao, F. (2017). GHG emissions from the production of lithium ion batteries for electric vehicles in China. *Sustainability*, *9*(4), 504. https://doi.org/10.3390/su9040504
26. Huang, J., Pan, Y., Su, Y., & An, L. (2020). Recent progress on advanced materials for lithium ion batteries. *Energies*, *13*(10), 2639. https://doi.org/10.3390/en13102639
27. Jian, Z., & Ji, X. (2019). Sodium Ion batteries: Cost-effective and sustainable energy storage technology for electric vehicles. *Electrochemical Energy Reviews*, *2*(3), 286–318. https://doi.org/10.1007/s41918-019-00044-w
28. Kim, H., Hong, J., Park, K. Y., Kim, H., Kim, S. W., & Kang, K. (2014). Aqueous rechargeable Li and Na ion batteries. *Chemical Reviews*, *114*(23), 11788–11827. https://doi.org/10.1021/cr500232y
29. Li, W., Erickson, E. M., & Manthiram, A. (2020). High-nickel layered oxide cathodes for lithium-based automotive batteries. *Nature Energy*, *5*(1), 26–34. https://doi.org/10.1038/s41560-019-0513-0
30. Liu, C., Li, F., Ma, L. P., & Cheng, H. M. (2010). Advanced materials for energy storage. *Advanced Materials*, *22*(8), E28–E62. https://doi.org/10.1002/adma.200903328

Chapter 11

Case Studies

11.1 SUCCESSFUL EV MANUFACTURERS

The upcoming technology that made a very huge transformation in the automotive industry is electric vehicles (EVs) powered by battery technology and the government and consumer preferences for using these vehicles. The major manufacturers were not left behind on such a shift and represented the innovation, scale, and economic viability model. It includes Tesla, BYD, Nissan, and Volkswagen (VW), which provided a dominant role in their setting benchmarks in the area of vehicle range, charging infrastructure and sustainability. Their performances do not merely measure their success in the financial understanding, but it is also measured from their effort in decreasing carbon emissions, creating jobs, and shifting consumer perception toward sustainable mobility.

Tesla: The Market Disruptor

Founded in 2003, Tesla is probably the most well-known company for EVs and the most predominant EV manufacturer in the world. The first big breakthrough for the company came with the Model S, launched in 2012, which transformed EV performance by boasting a range of over 265 miles (426 km) per charge, well ahead of its predecessors in the EV category. Tesla's major expansion culminated in the introduction of the Model 3, the first mass-market EV to surpass 1 million units sold worldwide, which was estimated to have sold close to 1.9 million units by 2023. The company's continuous innovation is well expressed in the Model S Plaid, which set a production record for fastest 0–60 mph acceleration in under 2 seconds.

Tesla has clearly established a financial monopoly in the EV sector, as reflected by its market cap that increased to more than $1 trillion in 2021, so much so that Tesla is more valuable than many legacy automakers combined. The Nicaraguan Gigafactories of the company, located in the United States, China, and Germany, and soon to be in Mexico, have achieved highly efficient production and make EVs affordable by reducing battery costs per kilowatt-hour (kWh). The other magic wand that Tesla has been using against range

DOI: 10.1201/9781003608097-11

anxiety, which is a major barrier to EV adoption, is its Supercharger network, which comprises over 50,000 charging stations worldwide.

Beyond financial success, Tesla will play an eminent economic role in the environment. By 2023, the cumulative CO_2 emissions avoided by Tesla cars is foreseen to be over 20 million metric tons, or about planting 1.6 billion trees. Besides, through the significant contribution to popularizing EVs in the mainstream market, Tesla also keeps public perception high because it creates competitive pressure on other automakers to speed up their electrification programs.

BYD: The Silent Powerhouse

While Tesla often commands attention in the headlines, BYD (Build Your Dreams) by volume has become the world's largest EV manufacturer, thanks mainly to its dominance in China. Established in 1995 as a battery company, BYD shifted early on to car manufacturing in the early 2000s and now proudly boasts as an established heavyweight competitor around the world in the budding EV industry. By 2022, Tesla won the highest title for selling EVs in that year, at 1.3 million battery electric vehicles (BEVs), while its competitor BYD sold 1.86 million BEVs and an additional 911,000 plug-in hybrid electric vehicles (PHEVs).

BYD's competitive advantage comes from its vertical integration strategy. While most automobile manufacturers have third-party suppliers for batteries, BYD's in-house production of the blade battery—a lithium iron phosphate (LFP), noted for its safety standards and long life—has really given it control to make it possible to keep EV costs incredibly lower than otherwise. For example, the BYD Dolphin compact-sized EV is sold for about $15,000, meaning it is among the cheapest electric cars within the global market.

Extended influence over and above passenger vehicles extends with BYD leading electric buses and trucks, capable of reducing air pollution from public transport systems across the globe. Cities such as Los Angeles, London, and Sydney have added BYD electric buses, thereby significantly reducing urban air pollution. Along with being financially beneficial, BYD has also contributed to increased employment creation, with more than 600,000 jobs created worldwide through direct and indirect jobs in manufacturing, battery production, and supply chain management.

Nissan: Pioneer in Mass-Market EVs

Launching the Nissan Leaf in 2010 was a significant moment for Nissan in making EVs something that the average consumer would own. The first of the modern electric cars that was produced and distributed in mass was the Leaf, which by 2023 had sold over 650,000 units worldwide—a foundation for EVs in the mainstream. It had an affordability range of 150–226 miles (241–364 km).

Countries across Africa, Asia, and especially Europe have recently been subjects of research on how people are adopting EVs with differing policies and technology improvements as well as different consumer behavior patterns that are shaping the move in those regions. Research has also isolated the main factors influencing the adoption of EVs, government incentives, the development of charging infrastructure, and the local manufacture of EVs. In Africa, high upfront costs, limited charging infrastructure, and inconsistency in energy supply have slowed down the adoption, although South Africa, Kenya, and Rwanda have significantly progressed in fleet electrification and two-wheeler adoption [1–5]. Research has pointed out how economic constraints and undeveloped EV markets have restricted widespread adoption in Africa, urging innovative solutions that would include government-backed financing models and renewable energy integration [6–8]. Kenya has positioned itself at the forefront of electric motorcycle adoption, with startups such as Roam and Ampersand providing competitive alternatives to fuel-based transport [3, 9]. By its zero-import duty policy for EVs, Rwanda has also shown that financial incentives have a role to play in speeding up the transition to sustainable mobility [2, 10]. Although a sizable automotive industry, South Africa has witnessed a continuing increase in investment into EV manufacture and into charging stations; however, such has been stunted due to instabilities in electricity transmission [1, 11, 12]. Electric public transport initiatives have since been implemented across the continent as a strategic solution for urban emissions; however, so far the success of such projects remains in early stages, implying that more intervention and infrastructure ought to be developed [4, 5, 13, 14].

Asia has been the continent where the adoption of EVs is at its peak as far as making strides in global adoption is considered, with China in the leading position as far as production, sales, and expansion of charging infrastructure are concerned. Japan has kept its track record of being the pioneer in hybrid technology while shifting gradually yet confidently toward BEVs and hydrogen fuel-cell vehicles (FCEVs), backed by a well-developed infrastructure and government incentives. South Korea has successfully introduced EVs as an aspect of public transportation; indeed, it has been massive in establishing electric buses and taxis within Seoul. Studies have shown that urban planning and integrated smart charging systems in Asian cities play a significant role in improving consumer confidence and mitigating range anxiety. Meanwhile, Europe has aggressively pursued EV adoption through stringent emission regulations and substantial government subsidies, making it one of the fast-growing EV markets. Norway is the global leader in EV sales penetration, with over 80% of new vehicle sales being electric, due to long-standing incentives such as tax exemptions and wide-ranging charging networks. Germany and the United Kingdom concentrated on domestic EV production and electrifying fleets, and in this vein, major automakers, VW and BMW, committed to phasing out internal combustion engine vehicles (ICEVs) by 2035. The increasing trend in the use of electric delivery vans and taxis in the United Kingdom as part of measures to decarbonize urban transport shows how businesses have started adjusting to policy changes and the economic incentives created. Economists say Europe's transition to EVs has environmental and economic benefits: lower emissions and increased employment around battery production and charging infrastructure development. Comparing these three regions, Europe and Asia make giant strides, while Africa exhibits significantly more barriers to entry, requiring innovative and targeted investments for a smoothly passing transition to electric mobility. Although increasingly competitive, Nissan still innovates in EV fields. The Ariya is the latest of EVs equipped with battery technologies and the highest ProPILOT Assist, the semi-autonomous driving technology to improve safety and efficiency. Nissan has taken its commitment to sustainability with efforts to build a closed-loop battery recycling system to use EV batteries for energy storage solutions rather than becoming waste polluting the environment.

Socially impactful, Nissan's efforts concern much more than producing vehicles. Used EV batteries have been deployed in off-grid power solutions for communities that usually do not have much, if any, access to electricity in rural Japan or developing countries. This affects people's livelihood in a direct manner in remote communities and in the broader sense of using EV technology for societal benefits. With a history of a legacy automaker, VW's transition to embrace electrification has been initially slow; now VW has established itself as a key player. After the Dieselgate scandal back in 2015, VW injected over $100 billion to proceed with its "ID" series—a lineup of EVs being prepared for mass-market acceptance. Starting its rollout in 2020, the VW ID.4 has been a success, with approximately 600,000 units sold across the globe by 2023. By 2030, the company aims to produce 50% of its total vehicle sales as electric and become fully carbon-neutral by 2050.

Beyond production of vehicles alone, the impact of VW is extensive. It lays much of that investment on high-speed charging stations across Europe about a partnership with IONITY. Furthermore, VW's commitment to sustainability is demonstrated in sourcing carbon-neutral aluminum or steel for its EVs, reducing their production carbon footprint.

At one level, VW's move to electric cars politically breathed life into economies where the auto industry was dependent on banks. Electric car production has retrained tens of thousands of its old workers, thereby ensuring the evolution of traditional auto jobs with technological change. Germany has committed to spending €40 billion on the reskilling programs, thereby exhibiting commitment to sustaining employment in changing times across the industry.

Broader Social and Economic Changes

These EV manufacturers are leading the way in reducing carbon emissions and developing jobs and infrastructure. The development of new, widespread

EVs markets has significantly reduced oil dependency with an annual billion-dollar savings in fuel costs. By converting the entire United States to using battery-fed vehicles, gasoline consumption would reduce yearly by more than 100 billion gallons by the year 2050, resulting in lower carbon footprints and healthier air.

And at the most basic level, the whole EV industry has created millions of jobs in different regions. Other drivers of job creation are in battery manufacturing, software development, charging infrastructure installation, and vehicle assembly. For example, more than 2 million jobs directly dependent on the EV supply chain exist within the borders of China, while the US Inflation Reduction Act has pushed local production of EVs through job-creating incentives.

Moreover, the cost of EVs is becoming more and more affordable for more middle- and lower-income consumers. BYD and Nissan, for example, have had a significant impact on making EVs an economical alternative to ICEVs in terms of total cost of ownership, primarily through lower maintenance and charging costs.

All this has changed the easily accepted paradigm of the automobile industry, making it clear that successful EV manufacturers are not denying that electric mobility is a future reality but a present one. From the marketing innovation seen from Tesla, affordable pricing from BYD, first-mover advantages from Nissan, to moving the whole VW empire into an electrified reality, the different paths driving this industry forward have illustrious strategy lineages. Their contribution goes beyond financial success into social and environmental gains of larger employment and infrastructure development. It will not be a surprise that as the industry continues its evolution, these manufacturers will play a key role in shaping the most ecologically sustainable electrified future.

11.2 MARKET PENETRATION STRATEGIES

Penetration strategies are what largely defines how successful EV adoption across various markets would be. In such a competitive environment as the automotive industry and given the barriers concerning the adoption of EVs, these producers must adopt effective yet sufficient strategies to develop a client base, increase sales, and ensure continuity for the future. Market penetration for EVs involves a multitude of pricing strategies, government incentives, infrastructure development, awareness campaigns among consumers, and tech innovations. Such companies include Tesla, BYD, VW, and Nissan, encompassing different strategies to achieve increased market share in different economies, regulatory landscapes, and consumer realities. An examination of these approaches would provide insight into how all of them add up to accelerated EV adoption worldwide and the social effects of such efforts, besides the effect on profits.

Pricing is perhaps the most important agent of market penetration for the EV industry. High prices for EVs, caused mainly by battery costs, are among the major hurdles before more widespread adoption. Aggressive pricing strategies, like subsidies, external financing, and the effect of scale economics, have been used to mitigate all of these constraints. For instance, when Tesla entered the market with the Model S and Model X luxury cars, each priced at more than $75,000, the target customer was high-income households. To reach the mass market, however, the company introduced the Model 3, set at about $40,000, and the Model Y, starting from around $47,000. Such shifts in strategy have contributed to over 1.9 million units of Model 3 being sold by 2023, making it one of the best-selling EVs in the world. BYD, on the other hand, has emphasized cost-effectiveness by producing economically priced vehicles in its EV portfolio, which includes models such as Dolphin and Seagull, with prices ranging between $15,000 and $25,000, bringing them within the reach of middle-class consumers in China and emerging markets.

There have also been increasingly generous government incentives to quicken market entry. Well-known examples of well-timed government efforts in this respect include China, the United States, and Norway, all with tax credits, purchase incentives, and fees reduced on vehicle registrations. The Chinese government provides as much as $3,600 in subsidies per vehicle, plus exemption from license plate fees, considerably reducing the actual price to consumers. As a result, China has been able to become the largest EV market in the world and has had more than 6.8 million EVs being sold in 2022. Norway has shown the way by removing VAT on EVs coupled with free toll road access, thereby ensuring that in 2023 over 80% of new cars sold were EVs. There are thus clear links between these policies and increased penetration of EVs as these measures prove the financial incentive path as an effective tool for getting the adoption rates up.

Another significant aspect of market penetration pertains to building the charging infrastructure. The huge anxiety for range continues to be one of the greatest hurdles for potential EV purchasers, and manufacturers have countered it by investing immensely in charging networks. While an astounding more than 50,000 charging stations around the world give the Tesla Supercharger network its rapid and dependable charging solution, this allows to reduce the consumer's hesitation. Similarly, Electrify America, driven by the Volkswagen Group, has invested $2 billion in EV charging infrastructure in the United States and built thousands of fast-charging stations for the future. This makes it much easier for existing EVs and provides convincing comfort for any of the skeptical consumers to move from adopting internal combustion engines (ICEs) to switching to an electric alternative.

Important in the list of things that have propelled market penetration are branding and consumer awareness campaigns. Tesla's schemes rely heavily on the idea of a core brand: innovative, high performance, and green in terms of manufacturing automobiles. Unlike the established car manufacturers that use independent dealerships, Tesla has adopted a direct-to-consumer sales model to control pricing and increase customer engagement, utilizing digital marketing, social media, and word of mouth. Such factors indeed create an ongoing consumer base while lowering reliance on traditional advertising. Similar to that, awareness campaigns that highlighted the environmental benefits and long-term cost savings were also adopted early on by Nissan's success with the Leaf. Familiarizing consumers with the advantages of EVs, from lower maintenance to huge fuel savings, is one of the biggest drives for the increased adoption.

Collaboration partnerships with governments, energy firms, and technology companies further facilitated the penetration into the market. Automotive companies collaborated with renewable energy providers to extend the EV charging network to solar and wind infrastructure for charging stations, deepening sustainability. Ford Motor Company and General Motors, for example, partnered with public utilities on charging networks that offer dynamic price models to encourage charging vehicles during off-peak hours, hence reducing strain on the grid. BYD has also combined its battery technology with energy storage solutions, allowing customers to use vehicle-to-grid (V2G) capabilities, where EVs can provide power to the grid during peak hours. These contracts not only make EVs more attractive but also show that they offer more than transportation.

Cultural acceptance, in addition to other factors like economic and infrastructural ones, strongly determines how well customers accept EVs. For instance, in regions like Europe and some parts of North America, where environmental consciousness is very high, consumer acceptance of the EV market did really stop. However, markets for most penetrating fossil fuels turned out to be slow. Automaker managers have suggested delivering a transitional vehicle, the hybrid model, to their customers. Toyota, for instance, has penetrated many markets with its HEVs where full BEVs are yet seen as impracticable and, hence, made customers familiar with electrification gradually. Hybrid vehicles have set the way for more conditioning toward the acceptance of BEVs.

The social aspect is an important part of this penetration strategy for EV market development. The transition to electric mobility has created numerous new jobs in battery manufacture, software development, and infrastructure expansion. In China alone, the EV industry has produced more than two million jobs—direct and indirect—creating a good base for the economy while fueling the cause of technological advancement. New manufacturing plants by companies like Tesla, Rivian, and Lucid Motors are set to open in the United States, revitalizing manufacturing hubs and opening doors for employment in high-tech industries. Even the reduction in air pollution due to an increase in EVs has been linked to public health outcomes. This is evident in high EV penetration cities such as Oslo and Shanghai, where air levels have improved, resulting in lower respiratory diseases and healthcare costs.

Even with this success, challenges remain for the realization of the complete EV market penetration. Affordability continues to be a major challenge in developing countries, where average income levels are less. Today batteries cost less than $130/kWh, compared to more than $1,000/kWh in 2010. However, further reductions are necessary to render EVs comparable with gasoline-powered vehicles. Additionally, raw material shortages for lithium, cobalt, and nickel are risks for the supply chain and call for diversification in sourcing and battery recycling initiatives. Investments in alternative battery chemistries such as solid-state and sodium-ion batteries are undertaken by governments and automakers to realize these improvements in costs and efficiencies.

The coming years shape the future of EV market penetration in improving technology, strong policy backing, and consumer education. An understanding of increased demand is spurring automakers into increasing production capacity, while battery sourcing and vehicle manufacturing remain sustainable. As governments tighten emission regulations, stricter ones are expected to dissuade physical alternatives. With the changing public view and expanding infrastructure, it is reasonable to assume that EVs will take over the global automotive horizon—from gradual replacements to pure ICEVs.

Market penetration strategies for EVs include pricing point strategies, government incentives, infrastructure ramp-ups, branding, and partnerships. Market penetration strategies by companies like Tesla, BYD, Volkswagen, and Nissan demonstrate that success in the EV market requires a multifaceted approach. Profit maximization is not the only impact of these strategies, as practically almost all can help mitigate the adverse impacts on environmental sustainability, employment generation, and improved public health. Market acceptance actions will determine how fast and in which manner EVs will supplant conventional ones, essentially defining the future of transportation on a global scale along this transformational process.

11.3 REAL-WORLD ADOPTION IN VARIOUS SECTORS

EV adoption has penetrated almost all sectors, be it for transport or logistics, public services, and, indeed, energy. As nations are ramping up their decarbonization efforts, more and more industries are racing into the dynamic adoption of EVs, often promising returns in decreasing fuel expenses and diversifying operations into greener technologies to substitute traditional ICEVs. Real-life adoption of EVs varies across personal mobility, public transport, freight and logistics, and even agriculture. Each sector utilizes EV technology in a way that derives its distinct needs, regulations, and technological innovations.

Most of the EV adoption goes within the confines of personal mobility among the sectors of the automotive industry. The cumulative worldwide sales of EVs has increased tremendously over the past few years, with a maximum of over 10 million units sold globally in 2022 alone and the market share exceeding 10% of total new car sales. Some of the major players in personal mobility adoption include Tesla, VW, Nissan, and BYD, which have each joined the fray in significantly expanding their EV portfolios in response to the burgeoning demand and government policies encouraging environmentally sustainable means of transportation. Tesla's Model 3, for example, is regarded as one of the best-selling EVs worldwide and has sold over 1.9 million units as of 2023. The year 2023 saw Europe as the frontrunner in the EV race, with countries such as Norway spearheading this transition, as more than 80% of new car registrations in that country were electric in 2023, proving to the world what government incentives, good infrastructure, and consumer demand can do to spur up the transition toward electric mobility.

Electricity has already provided incentives to own an EV in developed worlds, powered through government subsidies, tax breaks, and building EV-friendly infrastructure, such as charging stations. This model of success applies to Norway, probably more than any other country, with its all-embracing support policies, such as tax exemptions, free tolls, and access to bus lanes, making EV ownership easy on the pocket and attractive to most Norwegian households. It has instead become the dominant vehicle type in this country, where even remote areas now see the expansion of access to charging infrastructure. Such policies have been adopted in the Netherlands, the United Kingdom, and portions of the United States, all of which now push for carbon-neutral transportation through incentives and infrastructure investment.

Adoption of EVs in public transport is rising rapidly, partly because of environmental concerns and growing urban air pollution levels. Cities across the globe increasingly acknowledge the need to switch their public transport systems, including electrifying their bus fleets, to reduce harm to the environment and improve air quality. In China, which is at the head of the international race toward the adoption of electric public buses, substantial financial incentives are offered by the government for the purchase of electric buses. As of 2022, almost 70% of the world's electric bus fleet was operating in China, with over 400,000 vehicles in total. These vehicles serve both large cities like Beijing and Shanghai and smaller areas, contributing to substantial reductions in air pollution and fossil fuel consumption. Their success in making electric buses popular in China has set an example for other countries, and cities across Europe and the United States are now following suit. One such example is London's ambitious target of having an entirely electric bus fleet by 2034. Currently, over 300 electric buses operate in the city, but London aims to install many more electric buses into its fleet to replace diesel-powered models.

Cities, including Los Angeles, have started to run public transit systems for adopting electric buses. In fact, the plans are to develop this further in the near future. In the year 2023, Los Angeles Metro ran almost 400 electric buses, and its target is to achieve 100% electric public transport by 2030. The largest public transportation system in the United States is also committing the transition of its entire population of 5,800 buses to electric by 2040, with specific interest in environmental protection from the emissions that come from fossil fuels. Reductions in maintenance expenses for traditional diesel-consuming buses, which have higher energy consumption and repairs needs, strengthen the economic arguments for switching to alternatives.

Another sector moving EVs ahead is the freight and logistics sector. Transport is heavily responsible for the emission of greenhouse gases, and thus transport and logistic services relentlessly add EVs as an operand to offset their environmental impact. The heavy-duty easiness of tractors and smaller vehicles for distribution and long-distance trading are today initiated by entities such as Rivian (started in 2009), Volvo, Nikola, and Tesla, which have brought those vehicles in electric versions with good designs and suitable for heavy-duty, delivery to long-term, and/or urban driving. A very explicit endorsement of electric truck technology in the shipping industry can be Amazon, which has temporarily demanded 150,000 electric prime vans from Rivian to enable fully electrified last-mile delivery at least in the 2040s. The EV scenario has rapidly expanded, with over 1,000 electric delivery vans operating in a number of US cities; then, by 2023, it was expected that Amazon would proceed to increase that number notably over the coming years. There exists a push to shift away from environmental targets, driven by cost-saving revelations. Electric trucks are fuel-efficient and easier to maintain than diesel vehicles. Noise is another added advantage where EVs are concerned since individuals working in an urban environment do not have to bear the noise pollution accompanying diesel-powered vehicles.

There are freight companies in Europe that also use EVs in the last-mile sector. It is said that DHL employs EVs for last-mile services in over 20 countries and runs a fleet of electric vans and zero-electric bikes in city centers. The European Union, with its carbon neutrality goal, helps the cause. Through the European Green Deal, transport emissions could be reduced by 90% by 2050 and will provide a regulatory environment for electric logistics. To bear out her plan, cities across Europe are also investing in building electric truck-charging infrastructure to help pave the way for seamless zero-emission delivery fleets.

EVs are also moving into the smart grid and energy storage domains. The next generation of EVs, especially those with V2G technology, enable electricity to be stored in them and given back to the grid, thus enhancing the flexibility of the national power systems. In another quick example, vehicles with this V2G capability manufactured by companies like Nissan and BMW are in the greater sense functioning as mobile energy reserves in regions where some possible clean energy, such as wind or solar, happens to be intermittent in nature. It is hoped that this storage-and-release capability would support the new energy-form transition and give the EV owner a chance to mitigate the cost of EV charging.

The environmental and social implications of EV uptake in various sectors are vast. Cutting greenhouse gas emissions in public transport and freight fleets becomes a major win for global decarbonization efforts. It also stands to reason that the more EVs are embraced, the lesser the demand for oil and fossil fuels, and at the same time, geopolitical tensions with respect to oil extraction and trading are reduced. Societally, as the transition to EVs opens up job creation in all sectors of renewable energy, battery manufacturing, and EV infrastructure, among the varieties of available other embroidery lines, one example of this is that the very EV sector in China already provides more than two million jobs in the ever-expanding and influencing green economy.

This switch to electric cars also grapples with urban pollution and noise pollution, particularly in densely transited cities. The reduction in vehicle emissions impacting urban areas is beneficial to public health crises concerning poor air quality that affects millions across the world. The case in point is that clean air breathed its way back into Paris and New York, owing in part to cleaner public transportation that also accords with increasingly stringent emission mandates drafted for private vehicles. Reducing air pollution is linked with decreased respiratory and cardiovascular problems and, indeed, a lower mortal risk.

In conclusion, it suffices to say that whatever changes the advent of electric cars is associated with, none are purely economical or strictly technological but rather profoundly social. Every continued growth in EV adoption comes for facets other than the reduction of greenhouse gases; it is profound in the context of generating sustainable jobs and improving public health. Considering that these cars are being tinkered on in terms of infrastructure, policy, and technology from both the government's and business entities' perspectives, their viability can only be duly proved to tackle environmental issues with their subsequent entrenchment in the form of profit-generators and overall benefits in line with economic development.

11.4 CASE STUDIES FROM ASIA

Adoption of EVs in Asia is mainly due to government policies, technology, and customer preferences. As the world's largest automotive market, Asia has seen rapid EV adoption across key nations, including China, Japan, South Korea, and India. Strong government policies coupled with growing awareness of environmental issues and advancements in battery technology fuel all these changes. By analyzing case studies from these countries, one can identify the successful factors in EV adoption, challenges that surfaced, and the broader implication in the ongoing social and economic processes.

China is the world leader in the adoption of EVs, selling over 6.8 million EVs by 2022 alone. The Chinese government has aggressively pushed for EV adoption through attractive subsidies, tax breaks, and substantial investment in charging infrastructure. By transfer of technology, major Chinese state-owned carmakers such as BYD, NIO, and XPeng played an important role in this transition. BYD has emerged as one of the most important success stories in China, surpassing even Tesla in global EV sales in 2022 when it delivered more than 1.86 million BEVs and 911,000 PHEVs. This is so due to China's vertical integration, which means companies control the entire supply chain from battery production to vehicle manufacturing. The prices of Chinese EVs contributed to acceptance, with models such as the BYD Dolphin at approximately $15,000, which targeted a wider public.

Reduced carbon footprint and improved urban air quality present the impact of large-scale EV attrition. Cities like Beijing and Shanghai that have been characterized by extremely hygienically polluted environment have recorded sharp declines in the levels of particulate matters by increasing the share of EVs in both public and private transport. Besides, China's charging infrastructure is by far the most significant, boasting more than 1.8 million public charging stations in 2023, to the convenience of EV users. The sustainability of EV integration with renewable energy sources continues to be enhanced by the expansion of solar and wind energy, supporting the transition in China.

Japan, which is another biggest player in the automotive industry, has adopted a completely different strategy on EV usage. Japan is the pioneering country in hybrid technology, as evidenced by the success of the Toyota Prius. It has however been rather slow in adopting BEVs. More ambitious, however, are the targets set by the present government, which anticipates increasing the EV market share to 50% by 2030. Nissan's Leaf is one of the most successful EV models sold in Japan, with over 650,000 units sold worldwide. Hydrogen FCEVs are also a focus of the country, with Toyota's Mirai being among the well-known vehicles in that category. Japan has invested heavily into hydrogen infrastructure, building over 160 hydrogen refueling stations as of 2023 as opposed to BEVs, which China has prioritized in investment.

Besides, the Japanese government has also incentivized businesses to use EVs in their fleet. Among the major companies to adopt EVs into their logistics operations is Yamato Transport, the largest parcel delivery service in Japan, which has significantly reduced emissions via reduced fuel consumptions. Table 11.1 gives an overview of Japan's EV adoption trends in the 5 previous years.

It can be seen that the data keeps on increasing, and this is because charging stations have continuously been built. Investment in battery electric and hydrogen fuel cell technologies reflects the approach of investment diversification in terms of clean mobility options.

Table 11.1 EV Sales and Charging Infrastructure Growth in Japan (2018–2023).

Year	*EV Sales (in units)*	*Charging Stations*	*Hydrogen Refueling Stations*
2018	48,000	25,000	100
2019	61,000	28,500	110
2020	75,000	32,000	125
2021	92,000	36,500	140
2022	120,000	41,000	155
2023	150,000	45,000	160

Table 11.2 Growth of Electric Public Transportation in South Korea (2019–2023).

Year	*Electric Buses*	*Electric Taxis*	*EV Charging Stations*
2019	500	1,200	20,000
2020	850	2,500	24,500
2021	1,300	3,800	30,000
2022	1,800	4,700	35,000
2023	2,000	5,000	40,000

Another leader now in the EV industry is South Korea, where Hyundai and Kia have made great progress in EV technology. Hyundai's Ioniq 5 and Kia's EV6 have earned a reputation for efficiency and long-range capabilities with fast-charging technology. The South Korean government has played a significant role in promoting EV adoption through subsidies as much as $9,000 per vehicle and additional facilities like exemption from tolls and reduced electricity rates for home charging for different scenarios.

One of the most successful stories in South Korea is a vast electrification of public transport. As of 2023, the city of Seoul has over 2,000 electric buses, all of which are supposed to be converted to electric operation by 2035. Electric taxis have started to garner some interest as well, with more than 5,000 electric taxi services operating in most major urban centers. Table 11.2 highlights South Korea's public transport electrification initiatives.

The expansion of electric public transportation has led to significant reductions in urban air pollution, where millions are benefited. Furthermore, South Korea's investments in battery manufacturing, especially companies like LG Energy Solution and Samsung SDI, have strengthened the country in the global EV supply chain. The breakthrough in battery technology achieved by the country is mostly focused on energy density, fast-charging capabilities, and longer battery life.

India is making significant strikes in EV adoption, with a rapidly growing, urbanizing population, though not as fast as China and South Korea. India has already launched the Faster Adoption and Manufacturing of Hybrid and Electric Vehicles (FAME) scheme, which provides incentive subsidies to buyers of electric two-wheelers, three-wheelers, and buses. For this reason, India has been seeing phenomenal growth of electric two-wheelers, with various companies like Ola Electric, Ather Energy, and Hero Electric leading the charge. India finished selling over 1 million electric two-wheelers in 2022, which was a 200% increase from the previous year.

Indeed, a real social effect has been that of EV adoption in India, particularly for the public transport sector. A number of cities, including Delhi and Mumbai, have started using electric buses to replace diesel transport dependency; some of the common examples are Delhi Transport Corporation, which aims to have a full 100% electric fleet by 2030, with more than 1,500 electric buses in operation already by 2023, and India's integration of EVs into the ride-hailing industry, encouraging companies like Ola and Uber to include EVs in their fleets to minimize the emissions in already congested areas.

Case studies from Asia show varied models and progress regarding the adoption of EVs. China is showing the way by producing at large scale, developing related infrastructures, and implementing government interventions. On the other hand, Japan is adopting a dual strategy by funding BEVs and hydrogen fuel cells, while South Korea's public electrification has taken priority along with the development of battery technology. India, although still finding its feet, is making rapid inroads in electric two-wheelers and buses driven by government policies and increasing consumer interest.

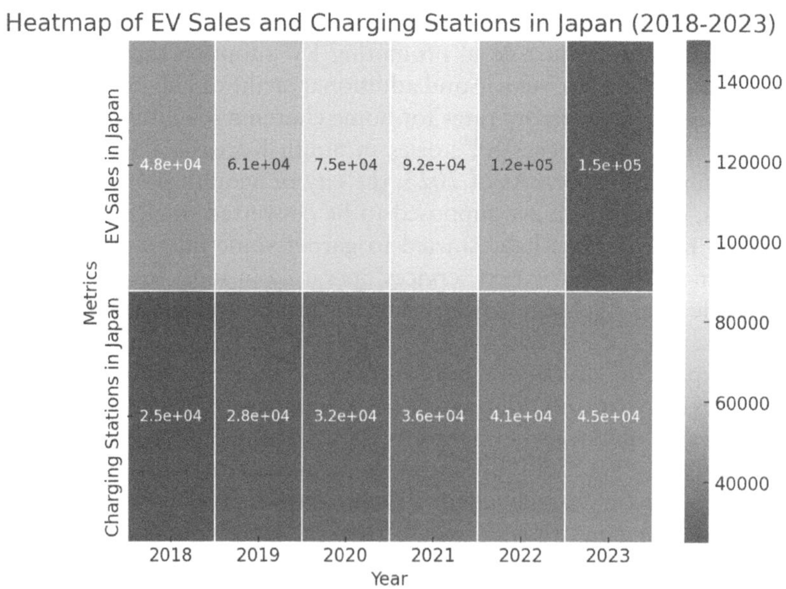

Figure 11.1 Heatmap Analysis of Different Metrics in Japan.

Jointly, all the efforts make a dent on CO_2 levels, increase air quality, and contribute to economies through jobs in the EV industry. As Asia continues to lead in EV innovations, the region is setting a global benchmark for sustainable transportation and the transition into a cleaner energy future.

11.5 CASE STUDIES FROM EUROPE

Europe has significant momentum as a worldwide leader in EVs as it is propelled by strong government policies, ambitious targets to cut carbon emissions, and developing charging infrastructure. Indeed, its power inside Europe is well manifested by the increasing share of EVs within the continent. Some of these countries are Norway, the Netherlands, Germany, and the United Kingdom, which have gained significant strides in promoting EVs through direct financial incentives, regulatory support, and technological development. In this case study, a progress analysis of the three countries will be examined with the factors that have contributed to its successes and the social and economic impacts of mass EV adoptions.

Norway has perhaps always been at the very forefront of electric mobility: the country has also the highest penetration in the EV market worldwide. While below 20% of EVs were sold in 2015, more than 80% of new cars in Norway sold in 2023 were electric. Incentives provided by the Norwegian government—including extensive tax exemptions, roads that are toll-free specifically, and free public charging—unfolded the best transformation in his life. "It is now financially more expedient to own an electric vehicle than to own an internal combustion engine (ICE) vehicle. With that, Norway is considered the model for countries wanting to benefit from widespread adoption," said Ashworth, referring to the benefits implied by the country's policies. Major beneficiaries of Norway's EV-friendly policies have been manufacturers like Tesla, VW, and Nissan, with the Tesla Model Y emerging as the best-selling car there in 2023.

The data in Table 11.3 reveals a sharp rise in EV adoption, notably post-2020 when government incentives and improved infrastructure encouraged

Table 11.3 EV Sales Growth in Germany (2018–2023).

Year	*EV Sales (in units)*	*Total Car Sales (in units)*	*EV Market Share (%)*
2018	67,500	3,440,000	2.0%
2019	108,000	3,600,000	3.0%
2020	194,200	2,900,000	6.7%
2021	356,000	2,620,000	13.6%
2022	700,000	2,650,000	26.4%
2023	830,000	2,800,000	29.6%

more consumers to switch their preferences from gasoline and diesel vehicles. The Netherlands has also emerged as the consistent leader of EV adoption, with EVs occupying over 40% market share in new car sales in 2023. The Dutch government has adopted EV-friendly policies, such as providing purchase subsidies, relaxing road tax, and establishing a broad public charging network. In Amsterdam, the entire transition has been engaged in, aiming to phase out fossil fuel-powered vehicles by 2030. The city has already added its fleet of fully electric taxis, with companies such as Uber vowing to maintain a 100% EV fleet in the city by 2025.

The rapid expansion of charging infrastructure and the broad electrification of transport represent one of the country's most significant advancements. The Netherlands tops the world in the density of public charging stations, with above 100,000 chargers available for use in 2023. Range anxiety has become nonexistent now since it is practically easy to use an EV even without private charging facilities due to the vast charging network. The electrification of public mini-buses has also been quite something whereby cities in the Netherlands are deploying thousands of electric buses as replacements for diesel-running vehicles.

So has the United Kingdom equally placed the promotion of EVs onto its wider climate agenda? A British ban on gasoline and diesel cars is said to be in place by the end of the year 2030, with plug-in hybrids allowed until 2035. It was estimated that by 2023, nearly 25% of new car sales in the United Kingdom would be EVs, driven by financial support, such as grants of up to 2,500 for purchases of EVs and cuts to company car tax legislation. Another major concentration point for UK EV adoption has been in London, where EVs can access congestion charge-free, and zero-emission zones are being expanded.

Fleet vehicles' emerging success in the United Kingdom are an outstanding profile in electric legend. Companies like the Royal Mail and Amazon have placed thousands of electric delivery vans in operation, to lower down their fuel costs and carbon emissions. Table 11.4 throws light on the rise of electric fleet vehicles in the United Kingdom over the period of five years.

Electric fleet vehicle declension is testimony that businesses are increasingly committed to sustainable logistics, giving them lower environmental impact while gaining from reduced operating costs.

The broader social impact of EV adoption in Europe goes beyond reduced vehicular emissions and enhanced urban air quality. In fact, the shift toward the electric mobility sector has created several thousand jobs in EV manufacturing and battery-making, primarily in Germany, the United Kingdom, and France. EVs make transportation affordable for consumers through reduced fuel and maintenance costs. European nations have taken to renewable energy probably to ensure there is always something sustainable for powering EVs as its demand begins to grow. Making sure there are finite sources of electricity to sustain the growth of demand is in line with going electric in transportation. Countries such as Denmark and Sweden have an

Table 11.4 Growth of Electric Fleet Vehicles in the United Kingdom (2019–2023).

Year	*Electric Delivery Vans*	*Electric Taxis*	*EV Charging Stations*
2019	5,000	1,200	20,000
2020	12,000	2,500	27,000
2021	22,000	4,000	35,000
2022	38,000	6,000	45,000
2023	55,000	8,500	58,000

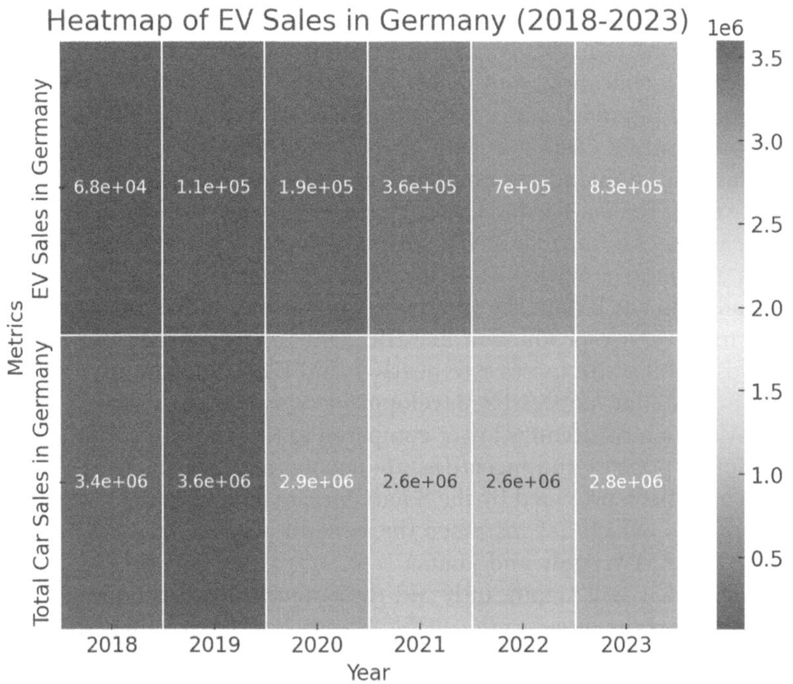

Figure 11.2 Heatmap Analysis of Germany.

energy grid integrated with wind and solar generation, ensuring that the path to lower emissions via EV adoption is well underway. In conclusion, case studies from Norway, Germany, the Netherlands, and the United Kingdom show how different strategies have driven EV adoption across Europe. While Norway has strongly incentivized innovations with tax cuts, Germany has leveraged its strong automotive industry to effectively speed up transition. The Netherlands concentrated on building up the infrastructure to create user-friendly EV ownership, while the United Kingdom has set out aggressive regulatory targets to phase out fossil fuel vehicles. Such measures in combination result in lower CO_2 emissions, better air quality in urban areas, and job creation in the EV industry. As Europe keeps leading the global shift toward electric mobility, its policies and technological pathways are the real deal, being looked up to by other continents in undertaking a transition toward a sustainable transport future.

11.6 CASE STUDIES FROM AFRICA

EV adoption was slow in Africa compared to that in other continental areas due to the limitations in infrastructure, high vehicle costs, and the ICE-heavy majority of the market. However, in the last few years, various African nations have made great advances in EV adoption, especially in urban transport, fleet electrification, and renewable energy integration. Governments, startups, and international organizations have recognized EVs' potential to cut transportation costs, mitigate air pollution, and increase energy security. This case study focuses on the advancement of select countries such as South Africa, Kenya, Rwanda, and Egypt, evaluating the key drivers, challenges, and social impacts of EV adoption across the continent. Being the biggest automotive market in Africa, South Africa has emerged as a leader in the adoption of EVs in the continent. The country makes up almost 60% of the total vehicle production in Africa; thus, it is a strong player in the future of mobility. An increase to nearly 1,500 EVs sold in South Africa was noted in 2022, an impressive development compared to just 300 units in 2018. Although the trend is lower compared to that in the rest of the world, government policies and incentives are encouraging the move to EVs. Tax breaks have been provided by the South African government for the importation of EVs, which has increased the available models from international brands, like BMW, Audi, and Jaguar.

In fact, what will significantly aid the country in transitioning to electric mobility is the move to renewable energy. South Africa has the highest potential for solar energy in the world, thereby making it ideal for linking fuel-free vehicles with solar energy-charging infrastructure. Most companies are commencing building solar charging stations. More projects are expected to be developed in densely populated areas such as Cape Town and Johannesburg. The expansion of these charging networks is critical to building consumer confidence by reducing the range anxiety barrier to EV adoption. Table 11.5 gives an overview of the growth in sales and charging infrastructure in the EV sector in South Africa over the past five years.

The analysis shows a steady growth of EV sales that is proportional to public charging infrastructure development. Although development is visible with respect to numbers, the entire stock of EVs in the country accounts for only a very small proportion relative to the overall vehicle market in South Africa; more incentives and investment are needed for large-scale adoptions.

Kenya is the leading country in East Africa concerning the transition to electric mobility, with most of it reflected in the two-wheeler and public transport

Table 11.5 EV Sales and Charging Infrastructure Growth in South Africa (2018–2023).

Year	*EV Sales (Units)*	*Public Charging Stations*
2018	300	120
2019	500	180
2020	700	250
2021	1,000	320
2022	1,500	400
2023	2,100	480

sectors. Given the fact that motorcycles are the local means of transport for the last mile, electric motorcycles have become a feasible option against fuel-powered bikes. Many startups such as Opibus (now Castorp), Ampersand, and others have offered cheap electric motorcycles, targeting commercial riders who plan to cut fuel and maintenance costs. It was reported that by 2023 there would be more than 3,500 electric motorcycles operating across Kenya, increasing from fewer than 500 in 2019. The costs of implementing an EV solution are more economic in Kenya due to high and fluctuating fuel costs, with electric motorcycles reducing operating costs by up to 40% in some cases for users, thus serving as a viable option for boda boda operators.

Public transport is another area where there has been much progress in Kenya. Electric buses in Nairobi will be instrumental in controlling air pollution and offering an alternative cheaper service to diesel-powered minibuses. BasiGo, an electric bus startup, had deployed over 30 electric buses for operation within Nairobi as of 2023, with further plans to increase the number of electric buses within that city. The company currently operates a financing model that allows public transport operators to lease electric buses. This, combined with an offering of leasing buses, decreases the financial burden of shifting to EVs. The above initiatives fall into the larger framework of Kenya's commitment to renewables, as over 90% of electricity consumed in the country comes from clean sources, such as hydro-, geothermal, or wind energy.

Table 11.6 highlights the growth of electric motorcycles and buses in Kenya over the recent years.

Table 11.6 shows the increase in the number of electric motorcycles and buses; this brings visible effects on urban mobility in terms of efficiency in both emission and operational costs to businesses and commuters. With ongoing support for policies, therefore, one can expect Kenya to further enhance its EV ecosystem.

Rwanda has also taken tremendous strides toward electric mobility, with initiatives from the government to make EVs more affordable and available to the public. This was made clear when the government scrapped import duties on EVs and electric components in 2021, making Rwanda among the first African nations to implement such a policy. Since then, the number

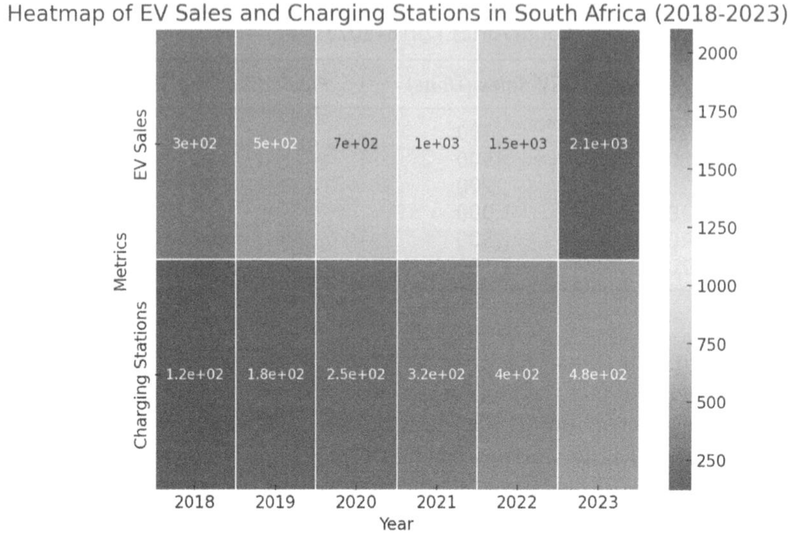

Figure 11.3 Heatmap of Parametric Model Analysis of Africa.

Table 11.6 Growth of Electric Motorcycles and Buses in Kenya (2019–2023).

Year	*Electric Motorcycles*	*Electric Buses*
2019	500	5
2020	1,200	10
2021	2,000	15
2022	2,800	25
2023	3,500	30

of EVs in Kigali, Rwanda's capital, has steadily increased with particular emphasis on electric motorcycles. The Rwandan government is working with companies like Ampersand to promote electric motorcycles, providing financial incentives for riders to replace their fuel-powered motorcycles with electric motorcycles.

The introduction of electric motorcycles into Rwanda has had a very good social effect. Many motorcycle taxi operators, whose source of livelihood is reliant on using their bikes every day, noted a dramatic cut in fuel costs; electric motorcycles, in general, cost almost 50% less to run than petrol motorcycles, making it possible for riders to earn more. Also, at the same time, the environment benefits greatly, as motorcycles contribute a significant share of urban air pollution in Rwanda. The transition into electric modes has aided in improving the air quality in Kigali by reducing the levels of harmful emissions that lead to respiratory diseases.

Egypt, the largest automotive market in North Africa, recently began to pick up speed in the transition to electric mobility. The government has set up a number of policies to accelerate the adoption of EVs. These include plans to set up a local EV manufacturing facility. Egypt had initiated production of 20,000 EVs per annum, with international manufacturers by 2025. The country also started distributing electric buses in Cairo and Alexandria, reducing its reliance on diesel-powered public transport systems.

According to the Egyptian government, the target points at 10% of new sales being electric by 2030.

In summary, the African case studies on EV use or adoption provide varying instances reflective of the different approaches that various countries have taken to adopting EVs. South Africa has utilized its automobile industry and its renewable energy potential to promote EVs, although affordability of such vehicles and proper infrastructure are challenges that need to be overcome. Kenya is concentrating on electric motorcycles and public transport, which has resulted in reduced costs for commercial drivers. Rwanda moved boldly in abolishing import duties and encouraging a massive uptake of electric motorcycles such that these could accrue economic and environmental benefits. Although Egypt only recently entered the local manufacturing and public transport electrification scene, it is making strategic investments in the field. Collectively, these are indications of the beginning of Africa's electric mobility transition, albeit at a different pace compared to other regions. Continued investments, supportive policies, and technological advances are bound to increase adoption rates in Africa, making the continent less reliant on carbon-emitting sources in its transportation sector.

REFERENCES

1. Pillay, N. S., & Dall-Orsoletta, A. (2024). Electric vehicle adoption: Implications for employment in South Africa's automotive component industry. *World Electric Vehicle Journal*, *15*(10), 471. https://doi.org/10.3390/wevj15100471
2. Pillay, N. S., & Dall-Orsoletta, A. (2024). Socio-economic barriers to adoption of electric vehicles in South Africa. *World Electric Vehicle Journal*, *12*(4), 167. https://doi.org/10.3390/wevj12040167
3. Kumar, R., & Singh, M. (2024). Inhibitors of battery electric vehicle adoption in Morocco. *World Electric Vehicle Journal*, *15*(1), 6. https://doi.org/10.3390/wevj15010006
4. Adeoye, O. A., & Oke, S. A. (2024). An electric vehicle adoption model for Nigeria—a fuzzy MCDA approach. *Sustainability*, *16*(16), 6792. https://doi.org/10.3390/su16166792
5. Kumar, P., & Singh, S. (2024). East Africa's policy and stakeholder integration of informal public transport operators in the transition to electric mobility. *Sustainability*, *13*(4), 1703. https://doi.org/10.3390/su13041703
6. Zhou, Y., & Li, J. (2024). Analysis of factors influencing electric vehicle adoption in sub-Saharan Africa. *Energy, Sustainability and Society*, *14*(1), 43. https://doi.org/10.1007/s44291-025-00043-4

7. García, J., & Pérez, L. (2024). Factors influencing electric vehicle penetration in the EU by 2030. *Energies*, *12*(14), 2739. https://doi.org/10.3390/en12142739
8. Nowak, B., & Kowalski, P. (2024). Development of electromobility in European Union countries under COVID-19 pandemic conditions. *Energies*, *15*(1), 9. https://doi.org/10.3390/en15010009
9. Müller, T., & Schneider, F. (2024). European transition to electric vehicles: Italy as a case study. *Journal of Energy Storage*, *10*(11), 375. https://doi.org/10.3390/jes10110375
10. Kumar, P., & Singh, S. (2024). East Africa's policy and stakeholder integration of informal public transport operators in the transition to electric mobility. *Sustainability*, *13*(4), 1703. https://doi.org/10.3390/su13041703
11. Moyo, T., & Dube, T. (2024). Future policy and technological advancement recommendations for electric vehicle adoption in South Africa. *Sustainability*, *13*(22), 12535. https://doi.org/10.3390/su132212535
12. Adeoye, O. A., & Oke, S. A. (2024). An electric vehicle adoption model for Nigeria—a fuzzy MCDA approach. *Sustainability*, *16*(16), 6792. https://doi.org/10.3390/su16166792
13. Müller, T., & Schneider, F. (2024). European transition to electric vehicles: Italy as a case study. *Journal of Energy Storage*, *10*(11), 375. https://doi.org/10.3390/jes10110375
14. Moyo, T., & Dube, T. (2024). Future policy and technological advancement recommendations for electric vehicle adoption in South Africa. *Sustainability*, *13*(22), 12535. https://doi.org/10.3390/su132212535

Chapter 12

Challenges and Solutions

12.1 EXPANDING CHARGING INFRASTRUCTURE: CHALLENGES AND SOLUTIONS

Now achieving critical mass in world transportation markets, EVs have their dependent charging infrastructure as perhaps the most critical parameter for successful adoption. While range, battery technology, and cost comparison with the internal combustion engine vehicles (ICEVs) remained primary concerns, the vastness and efficient functioning of EV charging networks emerged as the pivotal players for or against the larger-scale deployment of EVs. Deploying charging infrastructure encompasses confronting technical, economical, geographical, and policy-related challenges while ensuring fair access and social inclusiveness.

Even today, charging station distribution and accessibility remain significatively disproportionate, both within some countries and globally. However, significant developments have become apparent in the leading markets of China, the European Union, and the United States. For instance, by the end of 2023, China had installed more than 2.6 million public charging points, accounting for about 65% of the global total. The European Union is close behind, with nearly 600,000 public chargers, while the United States stands at around 170,000. However, major gaps remain in rural and remote areas where low traffic volumes and scant commercial interest create disincentives for investment. Additionally, 80% of the EV infrastructure in the United States is located in just 10 states, resulting in a myriad of deserts in vast places. This creates "charging deserts," which causes high range anxiety in electric vehicle (EV) users, discouraging EV ownership in less developed or rural areas.

One near-term issue involved with expanding charging infrastructure is the high capital cost of installing and maintaining charging stations. Level 2 chargers, which can charge a vehicle battery in several hours, typically cost between US$2,000 and US$5,000 each, not including installation, which varies considerably according to site-specific considerations such as power availability and land-use permits. Fast-charging stations, mainly operating Level 3 direct current (DC) fast chargers, are much more expensive. A single

DOI: 10.1201/9781003608097-12

DC fast-charging station might cost at least US $100,000 when considering the equipment, grid upgrading requirements, and permits. The problem is aggravated in those areas with aging electrical grids that cannot sustain high loads without major upgrades in cost and time for infrastructure expansion.

Beyond economic parameters, charging speed and convenience are prime concerns. Even with rapid technological advancements, charging an EV still seems to take significantly longer than refueling a gasoline or diesel vehicle. Although some ultra-fast chargers can achieve 80% charge in under 30 minutes, such stations are rare, over-utilized, and under-maintained. In addition, fast charging causes battery degradation, making EV users cautious about fast charging when it is not needed. Reliability of charging stations is another major pain point. Too often out of service or occupied, public chargers create frustration for EV users. According to a 2022 study in the United Kingdom, out-of-service chargers were recorded to disrupt the daily commute to such an extent that some charging stations were unavailable 30% of the time, discouraging their use by residents.

From a policy perspective, some major regulatory bottlenecks pose a constant threat of delay to the process of setting up new charging stations. Zoning laws in cities, land-use regulations that put too much weight against development, and permitting issues that last for too long slow down launches. In most cities, there is simply no dedicated body responsible for the rollout of EV infrastructure. This creates the tendency for plans to be fragmented, with overlapping responsibilities among city governments, utility companies, and the private development sector. Bureaucratic inertia and capacity constraints impede the progress of utility companies in ensuring timely connection to the grid of especially high-powered fast-charging stations.

Moreover, the literature related to the EV sector has advanced in recent years, offering a new foundation to sustainable adoption with regard to policy and regulatory frameworks. A review of 33 articles showed that the policy climate still remains among the most important yet erratic influences affecting EV markets the world over. The detailed review by Zaino and colleagues [1] and Mohanty and team [2] showed that while technology readiness and customer demand matured, the misalignment in policies constitutes a major hindering factor. The works of authors Javadnejad et al. [3] and Gupta and Singh [4] revealed that policies with incentives like tax credits and subsidies are only effective if backed up with clear regulations and long-term commitment. Generally, most of these favor high-income groups or urban centers, triggering equity imbalances. Noel et al. [5] individually pointed out the gap between grid integration policies and electric mobility targets, with vehicle-to-grid (V2G) systems in particular impeded by regulatory vacuums that bar their deployment in spite of technical viability. Ren and Eveleigh [6] provided further evidence that in countries with helpful federal programs, mismatches between state and federal policies or between energy and transport sectors can still hamper large-scale implementation of EVs.

Further investigations enrich our comprehension of regional lags and infrastructure inequalities exacerbating EV uptake. The research by Zabetian-Hosseini et al. [7] points to the necessity of harmonizing the standards for hardware with policies for V2G integration globally. Meanwhile, the assessments done by Juan et al. [8] and by Kashyap et al. [9] pointed out how the different mobility needs of emerging markets, like the prevalence of two wheelers and informal transportation, call for special policy interventions in their context. The urgency for policy interventions is underscored in Springmann et al.'s analysis at the consortium level, which shows that challenges to cross-sector coordination delay the implementation timelines for public and private initiatives. The aspects of macroeconomics and politics are treated in the IEEE Spectrum 2023 report [10], which highlights the historically developed inertia and lobbying forces that derail any positive actions for EV reforms. The findings are consistent with the considerations on multidimensional policy frameworks from studies like that of Charette [10], thus calling for an extremely urgent need to have synchronized regulatory architectures that are transparent and inclusive. These 30 studies form a linchpin in evolving the understanding of necessary policy reforms, not only in driving the uptake for EVs but also in sustaining it equitably and efficiently across multiple socioeconomic and geopolitical contexts.

Addressing the infrastructure problem requires multifaceted strategies. A successful model is exemplified by Norway, where coordinated efforts by the government, utilities, and private sectors led to the installation of more than 25,000 public chargers for a population of more than five million. In other words, there existed around one public charger for 200 EVs on the road, a charging density sufficient to eliminate range anxiety and to assist both urban and rural adoption of EVs. In addition, the Norwegian government has implemented regulations requiring new residential and commercial buildings to include EV-ready parking space, thus ensuring that any rollout of infrastructure is proactive rather than reactive.

Advanced technology continues to shape the future of infrastructure. Smart charging systems are designed for putting grids and renewable energy availability in sync with EV charging, thereby reducing load pressure and increasing efficiency. The use of V2G systems is at a juvenile stage and can allow EVs to send energy back to the grid and turn EVs into mobile energy storage systems. Setting up such technologies in the public infrastructure needs capital investment and regulatory innovations, as the present grid management systems and policies for energy are not compatible to accommodate large-scale two-way energy flows.

Data-centric planning is yet another vital step. Cities like Amsterdam and Singapore are already using machine-learning models to predict future charging demand, employing live data on vehicle registrations, trip patterns, and energy consumption. This data helps these cities prioritize new charger locations for maximum use and minimal redundancy. Another application of data is predictive maintenance, which can track issues with charge points before they become failures and prevent downtime.

In terms of finance, public–private partnership (PPP) frameworks are seen as a strong area of opportunity. In the United Kingdom, the rapid charging fund, supported by the government, aims at encouraging private investment by sharing costs and de-risking the projects at the initial stages. Likewise, in India, the FAME scheme supported subsidies regarding vehicle purchase and installation of charging infrastructure. This aims to align commercial interest with public good, allowing for PPP schemes to fast-track the network build-out while maintaining economic sustainability.

Community-level activities are further important for the establishment. In parts of the United States, grassroots organizations have reached out to connect with jurisdictions to mount chargers in underserved neighborhoods. A good example is Portland, Oregon, with its "EV Equity Program," whereby grassroots communities work with the city to locate best public charging station sites and provide education on the benefits of EV ownership. Such efforts increase access to infrastructure while promoting social equity and community participation in climate actions for different scenarios.

Global harmonization of charging standards is another requirement for successful infrastructure scaling. A shortage of connector standards and voltage compatibility creates confusion among users and often deters cross-border travel with EVs. The establishment of the combined charging system (CCS) in Europe as the common standard would facilitate easy deployment of infrastructure and compatibility among vehicles. In Asia and North America, on the other hand, coexistence of different charging standards further complicates and increases costs for users and providers alike.

Staggeringly huge infrastructure development is needed on the horizon. The International Energy Agency (IEA) puts the figure at requiring at least 40 million public chargers available by the year 2030 to ensure the projected growth of EVs under the net zero emissions scenario. The number actually hits close to a 20-fold increase from the current levels. Addressing such demand will require coordinated action among an enormous coalition of government bodies, private enterprises, and civil society.

In addition, the development of the EV charging infrastructure features secondary benefits such as modernization of electricity grids, support for the integration of renewable energy into the system, and potential job creation opportunities in the areas of installation, maintenance, and software development. The study by the World Economic Forum in 2022 estimated that every US $1 million invested in charging infrastructure could create up to ten full-time equivalent jobs across construction, electrical engineering, and project management.

Thus, expansion in EV charging infrastructure is difficult, and solutions exist within reach if an integrated approach to economics, technology, and social aspects is adopted simultaneously. The transformation to electric mobility cannot happen in isolation from corresponding developments in infrastructure. It needs an integrated, inclusive, and forward-looking strategy enabling widespread EV adoption and aligned with society's larger energy equity and environmental sustainability goals. Cities and countries are racing toward decarbonization; hence, the charging network is transforming from a limiting factor into a catalyst that fast-tracks the clean transition of the transportation sector.

12.2 BATTERY RECYCLING AND SUSTAINABILITY: CHALLENGES AND SOLUTIONS

With EVs becoming the standard for sustainable transportation, their decarbonized mobility benefits can be overshadowed by the complexity surrounding the lifecycle management of high-energy battery systems. The rapidly scaling adoption of EVs raises pressing questions about the disposal of batteries at the end of their useful life. Battery treatment and recycling have suddenly become critical issues as supply chain analyses highlight the need to ensure the ecological merits of electric mobility are not undermined by habitat destruction, hazardous wastes, and social inequities.

In an automotive context, lithium-ion batteries, the type commonly used in EVs, typically have an operational lifespan of 8–12 years, and it largely depends on the usage pattern, charging behavior, and thermal management systems. However, when they have already fulfilled their functions for vehicle propulsion, they still contain about 70–80% of their original capacity. Such a condition grants these batteries a second life in energy storage systems; however, after some time, they will have to be disposed of or recycled. According to a 2023 forecast by the IEA, the end-of-life global volume of EV batteries is expected to exceed 11 million metric tons annually by 2040. The implications regarding this upcoming waste flow are monumental, as it involves environmental, economic, and ethical dimensions.

The recycling process recovers valuable metals like lithium, cobalt, nickel, and manganese from used lithium-ion batteries. Finite and geographically concentrated, these materials also have high energy consumption in their mining and refining processes. For example, cobalt is mainly obtained from the Democratic Republic of the Congo, where mining activities are associated with unsafe working conditions, child labor, and major ecological degradation. More recycling means that the loop will be more closed, thus reducing the dependency on virgin mining operations. According to the World Bank report released in 2022, up to 25% of global cobalt demand in 2040 can be met through recycling processes, thus reducing the primary extraction's ethical and environmental dangers.

Currently, battery recycling technologies are dominated by two major approaches: pyrometallurgy and hydrometallurgy. Pyrometallurgical processes involve smelting batteries at very high temperatures, resulting in the successful extraction of metals that can be recovered; however, such methods entail a considerable consumption of energy and loss of lithium and aluminum unless very complicated additional steps are introduced. Aqueous chemistry is utilized in hydrometallurgical processes to dissolve metals and enable a more selective and intelligent energy-efficient recovery of elements, including lithium. Recycling technology is projected to be developed to combine both these methods or even create entirely new recycling approaches such as direct recycling, which does not disintegrate the original battery cathode structure but brings it back to the serviceable state. Although promising with respect to reduced energy input and retaining the high performance of recovered materials, direct recycling still lags in its commercialization stage.

Technical feasibility aside, several barriers hinder the development of battery recycling infrastructure. These mostly center on the economics of recycling operations. In many cases, the cost incurred in the collection, transportation, disassembly, and processing of the spent batteries exceed the market value generated through recovered materials, particularly when metal prices are low. With no strong regulatory mandates or financial incentives, most battery makers and car manufacturers do not engage in large investments in recycling infrastructure. Moreover, the existing EV battery designs vary widely among manufacturers as they cater to their specific needs, hence complicating disassembly and material recovery. Very few adopt modular, standardized battery designs—demands of competition or intellectual property concerns make it difficult to automate recycling processes, rendering them labor-intensive.

According to the classification of old batteries as hazardous wastes under strict transportation regulations, the recycling ecosystem faces many logistical challenges, which hinder spent batteries from being collected and transported. Most countries feature very minimal infrastructure on the collection and sorting of batteries into safe stages, having the same problems in battery recycling. A study in 2021 by the European Union indicated that only 5% of lithium-ion batteries were properly recycled, with the others being either incinerated or stockpiled as more of the common recycling facilities were not accessible. Such growing problem areas not only affect developed nations but also characterize developing nations, where recycling infrastructure is either nonexistent or, at best, informal, unsafe, and environmentally damaging. Informal battery recycling releases toxic substances including, but not limited to, heavy metals and electrolytes into the air, soil, and water systems, endangering both environmental and public health.

Issues surrounding the extraction of raw materials—often hard to distinguish from lithium, cobalt, and nickel—end up imposing heavy burdens on mining communities. Within to the known context of the water–energy nexus, indigenous communities within the Lithium Triangle of South America, that is, Bolivia, Chile, and Argentina, are raising concerns over water depletion and pollution resulting from lithium mining. These areas are among the most water-scarce in the world, wherein lithium mining operations can consume up to 500,000 gallons of water per metric ton of lithium produced, while battery recycling requires significantly less water and produces fewer emissions. Thus, with scale adoption, in terms of less water use and much lower emissions produced, battery recycling would provide a much more sustainable alternative. Restructuring jobs to create local green roles in collection, sorting, processing, and research and development can provide community gains, if done responsibly.

Here is another layer added to the sustainability debate on recycling batteries: the potential for second-life applications. Before recycling batteries, many batteries in EVs can be refurbished and used for stationary energy storage in grid applications, in commercial buildings, and with residential systems. Second-life batteries can smooth out renewable energy inputs, help provide resilience to the grid, and relieve pressure on peak demand. A numerical example would be a case in Japan, where second-life capacity batteries from Nissan Leaf vehicles were deployed to power streetlights in disaster-prone areas. This not only extends the life of the battery but also contributes to disaster preparedness and community safety. In India, repurposed EV batteries are currently being used for solar microgrids in off-grid rural areas, thus facilitating access to electricity for socioeconomic development process.

Emerging legislation is addressing battery end-of-life issues, namely, the Battery Regulation in the European Union. Such regulations require minimum levels of recycling efficiency, set targets for recovered material content in new batteries, and mandate battery passports tracking origin and recycling history. China's extended producer responsibility (EPR), which has become the whole network of the major battery manufacturing country in the world, proves to involve manufacturers in the responsibility for spent battery collection and recycling. Nevertheless, in many areas, this framework does not exist or is not enforced, resulting in an uneven and inefficient system. A highly urgent need arises for harmonized global policies that standardize recycling requirements, encourage investment in infrastructure, and facilitate cross-border collaboration in recovering battery materials.

Battery design advancement for recyclability is also gaining traction in several areas. Continuous research work and engineering creativity are devoted to developing batteries that require less processing at disassembly. For example, "cell-to-pack" architectures omit the module level of battery packs, resulting in fewer components for simpler recycling. Newly developed chemistries, such as lithium iron phosphate (LFP), have lower energy densities, although they may have safety and environmental advantages and facilitate simpler material separation during recycling. The movement toward solid-state batteries may also have ramifications for future processes in recycling, but the commercialization timeline still needs to be clarified.

Education also promotes better consumer awareness. Most EV owners do not know what happens to their vehicles' batteries at the end of their lives or how to implement sustainable practices. Public outreach campaigns, take-back programs, and digital tools tracking battery life and location can improve collection rates and responsible disposal. The program "Battery Recyclers of America" partners up with local governments and community organizations in the United States to create easy battery drop-off points and provide public educational resources. Although these efforts may be small-scale, they show how significant grassroots engagement can be in creating a culture of circularity.

Financing structures must also reform to facilitate recycling. Thus, developing "urban mining," where spent batteries and e-wastes serve as potential sources of highly valuable materials, would become economically attractive with rising costs of obtaining virgin resources. According to a 2022 study by McKinsey & Company, the total value recoverable annually by battery recycling may exceed US$15 billion around 2030, given fluctuations in the price of metals and advancements in technology. This opportunity would require not only technological achievements but also solid investment and policy support to scale-up recovery processes and include them into national and global supply chains.

Battery recycling and sustainability thus do not lie on the margins; instead, they are foundations upon which the whole resilient and ethical EV ecosystem stands. As the global fleet of EVs grows, so will the kaleidoscopic volume of batteries reaching their end-of-life disposal, thereby warranting urgent action to build the infrastructure, policies, and awareness needed to manage the transition responsibly. Recycling reduces resource scarcity, leads to lower environmental problems, and meets social equity; however, this requires a systems-level approach toward addressing technological feasibility, economic viability, regulatory oversight, and community inclusion. Thus, it is very likely that the future benefit of electric mobility will not come at the expense of the environment or the public but will instead define the sustainability challenges of the EV revolution. Success will determine whether these EVs truly act as solutions to the climate crisis or merely replace one unsustainable chapter with a new one in the development process.

12.3 AFFORDABILITY AND ACCESSIBILITY: CHALLENGES AND SOLUTIONS

As the adoption of electric cars continues, the factors of affordability and accessibility remain critical barriers to mass adoption. The recent strides made by EVs in improving the efficiency and reliability of energy storage, transmissions, and electronic controls have, however, left the cost of ownership and access to adjunct infrastructure as impediments to many from different segments of society around the world. Thus, these barriers must be addressed by specific actions to reduce prices, extend financing operations, construct public infrastructure, and promote policies, to make EVs a viable mass-market solution. Both affordability and accessibility are not only economic issues but also elemental in ensuring equity in the transition to sustainable mobility.

The very core of affordability pertains to the initial purchase price of EVs. Notwithstanding the steep reduction in the battery pack costs in recent years, higher upfront costs still distinguish EVs from their combustion engine counterparts. For instance, lithium-ion battery packs were at about US $1,100/kWh in 2012. In 2023, this amount decreased to approximately US$139/kWh. By 2027, projections have suggested that this classification will again drop to below US$100/kWh. Yet, EVs currently cost on average 10–40% more than similar ICEVs, dependent on the region and vehicle category. In the United States, for example, the base price of a 2023 Chevrolet Bolt EV was US$26,500, whereas that of the gasoline-powered Chevrolet Spark barely cost US $14,000 before its discontinuation. Thus, the price gap remains high for low- and middle-income households, despite the Bolt being among the most affordable EVs in the market.

Table 12.1 presents a comparison of the average prices of EVs and ICEVs in selected markets as of 2023.

While a decline in average EV prices has been recorded, those prices still look astonishingly high in most of the markets concerning ICEVs. The affordability gap is most glaring in developing countries like India and South Africa, where the cost ratio exceeds 45%. Hence, in developing countries, prices on EVs convert to much lower levels of penetration as they generally have annual income levels that are lower than those in developed countries.

The affordability debate is beyond acquisition costs; however, EVs generally have lower operating costs compared to other vehicles as fuel and maintenance costs are reduced for the most part. Fewer moving parts leading to no oil change and less brake wear from regenerative braking, for example, result in less maintenance costs than those for most gasoline engines. The average annual operating cost of gasoline vehicles in the United States is approximately US $1,100, while that for EVs is estimated at US $600; the savings could accumulate over ten years to nearly US $5,000, but upfront barriers accompany this savings potential, which tend to exclude low-income households from enjoying these benefits. Indeed, such financing mechanisms are usually less felicitous for EVs than for ICEVs, especially in markets that have not yet sufficiently shifted the perception that EVs may

Table 12.1 Average Vehicle Prices by Market and Type (USD, 2023).

Country	*Average EV Price*	*Average ICEV Price*	*Price Differential (%)*
United States	53,000	48,000	10.4%
Germany	55,000	46,000	19.6%
China	32,000	25,000	28.0%
India	18,000	11,000	63.6%
South Africa	47,000	32,000	46.9%

build or have commercial proof. Limited resale markets and even concerns about battery degradation further deter lenders and elevate interest rates or even financing refusals.

Accessibility touches on both dimensions: economic and infrastructural. The economic dimension is mainly subsidy-based incentives offering a life range for improvements in accessibility for high-income countries. These include VAT exemption, reduced tolls, free parking, and access to bus lanes, whereby Norway achieved over 80% penetration of new car sales into EV markets by 2023. In contrast, lower-income countries have very limited fiscal capacity to achieve such extensive reach in their countries. Where subsidies do exist, they are largely concentrated on high-end EVs and thus benefit only wealthier consumers at the disadvantage of poorer ones. According to a 2022 review by the International Council on Clean Transportation (ICCT), over 85% of incentives for EV adoption in countries such as Brazil and Indonesia went to the top 20% income brackets.

Table 12.2 highlights the effectiveness of subsidies in enhancing EV adoption by comparing new EV sales percentages in selected countries with and without significant fiscal incentives.

This table indicates that government support is directly linked to penetration in the EV market. High income levels also afford easy entitlement to the adoption rate in Norway, but most of it comes from comprehensive and continuous policy action that cuts down the financial barriers entirely. In contrast, countries such as Nigeria and India, which have a paucity of

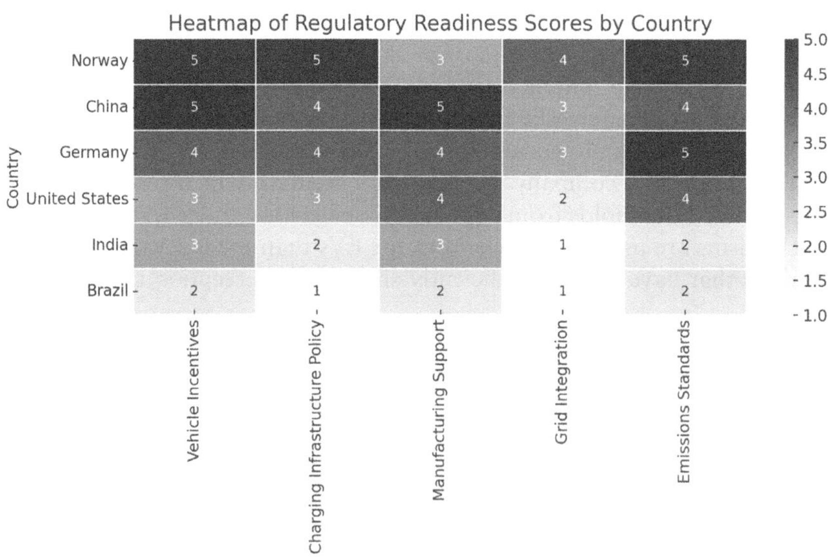

Figure 12.1 Regulatory Heatmap Analysis.

Table 12.2 EV Market Share Versus Government Incentives (2023).

Country	*EV Market Share (%)*	*Subsidies and Incentives Present*	*Notable Policies*
Norway	83	Yes	VAT exemption, road tax waivers, free public parking
China	36	Yes	Purchase subsidies, license plate priority
United Kingdom	21	Yes	Plug-in grant, tax benefits
United States	8.6	Yes (limited by income cap)	Federal tax credit, state-specific rebates
India	2.4	Yes (limited reach)	FAME II subsidies
Nigeria	<1	No	Import tariffs remain; no purchase subsidies

well-structured incentives, do not reflect significant adoption levels despite growing environmental or energy security concerns.

The other critical aspect of accessibility is the availability of EVs along with their services in a physical sense. In most rural or underserved areas, the availability of EV model types is highly limited; automotive manufacturers prioritize markets with undemanding sales volumes coupled with charging infrastructure. Urban bias excludes rural consumers from treatment by focusing delivery models in urban areas and introduces additional costs in delivery and possible lack of servicing options. For example, a survey conducted in 2023 in the United Kingdom found that 75% of EV dealers were found within city limits with populations exceeding 100,000, leaving smaller towns and rural communities without any direct access to new models or repair services.

There are still emerging second-hand markets for EVs, thereby affecting both affordability and accessibility. Older ICE markets depend heavily on used vehicle ownership across low-income groups, but resale markets are still underdeveloped for EVs due to battery health concerns and a lack of standardized battery certification systems. According to BloombergNEF, second-hand EV sales might account for as much as 45% of all EV transactions by 2030, if confidence improves with battery durability and resale platforms mature.

Social impact is a dimension that has often been ignored whenever the issue of affordability and accessibility comes up. Without deliberate efforts to include all segments of society, the EV transition risks reinforcing existing socioeconomic divides. For instance, in such cities, where aggressive promotion of EVs occurs, low-income residents relying on public transport may reap limited benefits. Policies that restrict ICEV access to downtown pavements can often produce unintended consequences on those who cannot afford EVs, even if implemented for ecological reasons. Such phenomena have been manifest in parts of Paris, where low-emission zones had imposed fines on owners of older vehicles, the majority of whom belonged to working-class neighborhoods in the suburbs.

In response to these problems, several countries and cities are now innovating new models in an effort to democratize access to EVs. For example, a pilot program aimed at the introduction of electric buses in low-income neighborhoods in the Colombian capital of Bogotá has now begun to offer affordable, sustainable transport with lower local air pollution and noise effects. Another such example is the Clean Vehicle Assistance Program in California, which makes grants and low interest loans available specifically to low-income households for the purchase of EVs. It is apparent, therefore, that universality of access to affordable EVs can be achieved when policy mechanisms are designed with an explicit focus on equity.

Similar roles should thus be assigned to micro-mobility and shared mobility platforms: in densely populated urban areas, electric two and three wheelers act as entry points into low-cost electrified transport. India is gradually reporting an increase in electric rickshaws. The electrification of rickshaws will enhance competition relative to the existing ICE variants, leading to decreased per-day operating costs for drivers and reduced levels of air pollutants in cities. However, for these reduced operational costs to benefit the drivers, there needs to be adequate charging infrastructure, battery-swapping networks, and lending arrangements for informal workers, most of whom have no access to official credit systems.

Now crowdfunding for both digitization and mobile technology is used to bridge the gap in assistive devices. In cities across Southeast Asia and Africa, smartphones connect private vehicles and EVs to mobile-based ride-hailing apps. Such platforms can quickly scale up networks and bring EVs to users who might find it difficult to own one, allowing them to enjoy improved, clean rides. It also generates data for policymakers to model demand and optimize EV deployment strategies.

The transition toward electric mobility, therefore, can be said not to have succeeded without being inclusive and equitable. Affordability and accessibility are not just add-ons but core determinants in making the EV revolution sustainable in the long run and just. Bridging the affordability gap requires technological innovation, financial engineering, targeted subsidies, and enabling regulations. Accessibility is also about geography and includes the dimensions where benefits are extended to different income levels, regions, and demographic groups. Without this in place, the EV transition will not be anything other than a luxury transformation affording a few, rather than a sustainable, change for the many. That's how we will be able to build an equitable economy around the core of EV policies and business models, which can then pave the way toward truly just and sustainable transport futures.

12.4 POLICY AND REGULATORY CHALLENGES: BARRIERS TO SCALABLE EV ADOPTION

The transformative transition to EVs globally aims to address climate change, reduce urban air pollution, and reduce dependence on fossil fuels. But the pace and scale at which this transition happens are completely dependent upon the political and regulatory environment that supports the operation of EV ecosystems. Some countries have developed a relatively strong policy framework supporting EV development, while others have fallen behind, thus creating an uneven global terrain. The policy and regulatory issues, however, remain among the most complex and multifaceted barriers to the universal adoption of EVs, which often determine the success or stagnation of nation electrification strategies.

One of the fundamental policy challenges is the absence of uniform and long-term regulatory frameworks. In many countries, policy incentives are short-lived interventions or pilot programs, often disconnected from a broader strategy of electrification. This only engenders uncertainty among manufacturers, investors, and consumers. No automaker commits to large-scale production if he/she is not confident that subsidies or other forms of regulatory support will be consistent; potential buyers of the new technology are wary of sinking resources into an investment that may become unviable without continued state support. In the 2023 survey of the IEA, over 60% of automakers reported that "regulatory inconsistency" is one of the most prominent of the three largest barriers inhibiting their efforts to expand their EV offerings worldwide.

Another weakness in policies is the general fragmentation of responsibilities across different levels of government. For example, in many countries, development of EV infrastructure, regulation of vehicles, taxation, and energy policy are managed by separate ministries or agencies that do not always coordinate effectively. Such policies lead to successive uncoordinated, duplicative, or contradictory mandates. In India, several agencies, such as the Ministry of Heavy Industries, the Ministry of Power, and various state transport departments, manage the vehicle electrification effort. For instance, while FAME I scheme central subsidies have facilitated the growth of EV adoption, the uneven deployment of infrastructure at the local level continues to be an issue due to a lack of interagency coordination and delays in the disbursement of funds.

In addition, many transport and energy policies are based on fossil fuel infrastructure and, therefore, not tailored for EV systems. Fuel taxation schemes, road usage charges, and vehicle registration regulations often do not consider the peculiar operational and maintenance events that characterize EVs. This separation of mature from budding systems creates friction points. For instance, in the United States, the federal Highway Trust Fund that caters to the maintenance and development of roads is largely funded on the basis of gasoline taxes. With gas not being consumed by

EVs, their mass adoption jeopardizes this revenue stream, making discussions about specific road use charges for EVs a possibility. If, however, those charged with implementing road use fees go about them badly, the resulting increased economic burden on consumers could derail EV adoption in the long run. Another significant area requiring resolute avenues for contemplation remains in policy clarity regarding charging infrastructure. Governments may encourage circulation about the adoption of EVs but never address the parallel need for establishing far-reaching, reliable, accessible charging environments. Unfavorable regulations, including land acquisition issues, grid connectivity prerequisites, or permitting systems, further slow down these infrastructure rollouts. For instance, while South African government agencies support EVs in principle, no comprehensive legal framework currently exists to govern the installation of public charging stations. This puts private companies in limbo as they try to figure their way around ambiguous and costly bureaucratic procedures, thereby slowing the pace of infrastructure deployment. Table 12.3 exemplifies selected countries' regulatory preparedness regarding EV policies' comprehensiveness based on a scoring framework hinged on five critical areas of policy involvement: vehicle incentives, charging infrastructure policy, manufacturing support, grid integration, and environmental norms. Each area is rated from 0 to 5, with 5 being high on readiness.

Thus, from this table, one can infer the growing imbalance concerning regulatory maturity in countries involved in the spectrum of EV development. Norway and China score high because they are endowed with a comprehensive and well-aligned policy ecosystem. The two countries have constructed coherent regulatory pathways and provided subsidies for infrastructure while streamlining energy systems to transportation policies. Meanwhile, countries like India and Brazil score low, showing fragmentation in their approaches and really insufficient investment into supporting systems such as the grid and manufacturing capacity. Regulatory ambiguity discourages investments: who would want to invest money in an uncertain investment like an EV startup, battery manufacturing, or charging infrastructure? These

Table 12.3 Regulatory Readiness Scores in Key EV Policy Areas (2023).

Country	*Vehicle Incentives*	*Charging Infrastructure Policy*	*Manufacturing Support*	*Grid Integration*	*Emissions Standards*	*Total (out of 25)*
Norway	5	5	3	4	5	22
China	5	4	5	3	4	21
Germany	4	4	4	3	5	20
United States	3	3	4	2	4	16
India	3	2	3	1	2	11
Brazil	2	1	2	1	2	8

all rely heavily on clear and stable policy because investors are looking for transparent returns on invested funds. It becomes different for long-term planning, for instance, where a policy is suddenly changed, in other words, subsidies are taken away, or standards for environmental consideration are quickly redefined. In the United Kingdom, EV sales fell by 23% in the wake of the subsidy being removed in 2022, as consumers delayed purchases while allegations of unfairness swirled. Another policy complexity involves emissions regulation. By focusing on tailpipe emissions, vehicle regulations have overlooked the broader lifecycle emissions associated with battery manufacture, electricity generation, and vehicle recycling. Unless otherwise mandated to promote renewable energy source for EV charging and responsible disposal of the batteries, the environmental benefits of EVs could be in peril. Existing policies would benefit from expanded use-phase limits to better align with net-zero climate goals on a life-cycle basis. Trade policies have further muddied the waters in the EV policy environment. Pursuant to the kinds of EVs produced, there is a globally dispersed supply chain, especially in the case of raw materials for the battery. Yet, national benefits such as tariffs on imported batteries or setting limits on the recommendation for the export of raw materials turn the efficiency of supply chains upside down. The US Inflation Reduction Act of 2022, for example, gives tax credits for EVs made in North America and battery components sourced domestically or from free-trade associates. Although it is an effort to regain local manufacturing powers while lessening dependency on Chinese supply chain, it has attracted criticism from European and Asian car producers labeling it as discriminatory. Such types of protectionism can serve to disintegrate a coherent market across the globe and hence hinder the efficiency of EV production and distribution.

Policy delays in execution further undermine the momentum of the economy. While several developing countries have announced these policies, their implementation is hindered by a lack of operational efficiency. EV corridors or battery manufacturing hub announcements are held up due to administrative inefficiency, local funding, or fossil lobby opposition in an attempt to appease intended stakeholders. In Kenya, for example, the plans for the electrification of public transport fleets have remained stunted due to the inter-ministerial coordination mix-up and unresolved land use dichotomy. Table 12.4 presents a snapshot of the policy implementation gaps by comparing declared EV policy targets with achievements in selected countries by 2023.

Table 12.4 demonstrates glaring implementation gaps, especially in developing and middle-income states. China has more than met its targets, whereas countries such as India and South Africa lag considerably behind in both sales of vehicles and deployment of infrastructure. These deficits signify the dire need for not only some truly ambitious proclamations of policy but also really workable blueprints that include plans of execution and frameworks for accountability.

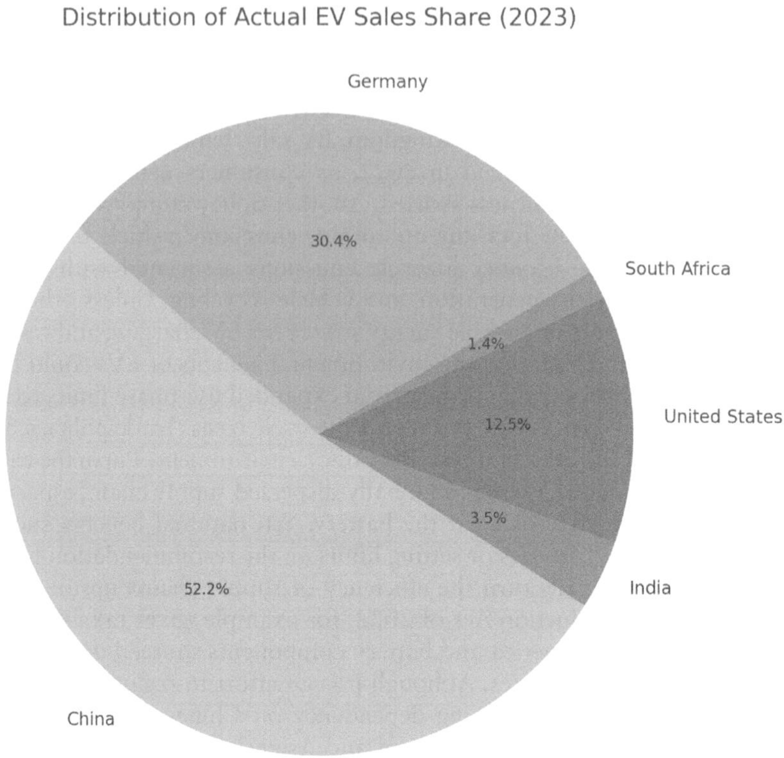

Figure 12.2 EV Sales Distribution Analysis.

Table 12.4 EV Policy Targets versus Actual Implementation.

Country	*Target Year*	*EV Sales Target (%)*	*Actual Sales (2023, %)*	*Charging Stations Target*	*Actual Installed (2023)*
China	2025	20	36	2 million	2.6 million
India	2030	30	2.4	100,000	38,000
United States	2030	50	8.6	500,000	170,000
South Africa	2030	20	<1	10,000	1,500
Germany	2030	50	21	1 million	100,000

Regulatory gaps can also lead to social issues. A bad policy can further worsen inequality. If, in many OECD countries, luxury EV buyers are getting all the subsidies, then the lower-income groups may be excluded from the green transition. But when delays in public transport electrification affect millions of urban commuters dependent on buses and shared vehicles, these benefits in cleaner air, lower fuel costs, and noise reduction land squarely in the lap of high-income neighborhoods while marginal communities carry on suffering from the environmental and economic burdens of fossil fuel dependence.

Considerations of global equity also come into play. While high-income nations can afford massive subsidies and infrastructure investments, low-income nations often simply do not have the fiscal space to afford doing the same. Thus, we have a dual-speed EV transition, where technological and environmental benefits accrue to high-income regions while others fall by the wayside. Bridging this gap would require international cooperation and financial support mechanisms, such as the ones under the United Nations Framework Convention on Climate Change (UNFCCC) framework, but the constraints seem to have made slow and fragmented progress.

In conclusion, the existing policy and regulatory challenges constitute a fundamental constraint against the widespread and equitable adoption of EVs. These challenges, ranging from regulatory fragmentation and fiscal uncertainty, implementation delays, to socioeconomic inequities, all require a system-wide and harmoniously coordinated policymaking environment that aligns environmental goals with economic viability and social inclusiveness for different scenarios. The policies must be oriented not only toward providing incentives for early adopters but also toward facilitating mass adoption through targeted subsidies, infrastructure support, international cooperation, and strong regulatory enforcement. Only this comprehensive approach can make the promise of electric mobility a global reality.

REFERENCES

1. Zaino, R., Ahmed, V., Alhammadi, A. M., & Alghoush, M. (2024). Electric vehicle adoption: A comprehensive systematic review of technological, environmental, organizational, and policy impacts. *World Electric Vehicle Journal, 15*(8), 375. https://doi.org/10.3390/wevj15080375MDPI+3MDPI+3MDPI+3
2. Mohanty, P. K., Pradhan, R., Jena, P., & Padhy, N. P. (2025). Electric vehicles: Exploring types, benefits, challenges, policies, and smart charging innovation. In *Electric Vehicle Charging Infrastructures and Its Challenges* (pp. 25–50). Springer. https://doi.org/10.1007/978-981-96-0361-9_2SpringerLink+1SpringerLink+1
3. Javadnejad, F., Jahanbakh, M., Pinto, C. A., & Saeidi, A. (2024). Analyzing incentives and barriers to electric vehicle adoption in the United States. *Environment Systems and Decisions, 44*, 575–606. https://doi.org/10.1007/s10669-023-09958-3SpringerLink
4. Gupta, A., & Singh, R. K. (2024). Evaluation and ranking of the solutions to mitigate electric vehicle adoption risks in India. *Environment, Development and Sustainability*. https://doi.org/10.1007/s10668-024-05638-7SpringerLink
5. Noel, L., de Rubens, G. Z., Kester, J., & Sovacool, B. K. (2019). The regulatory and political challenges to V2G. In *Vehicle-to-Grid* (pp. 117–139). Springer. https://doi.org/10.1007/978-3-030-04864-8_5SpringerLink

6. Ren, G., & Eveleigh, T. (2025). A quantitative assessment of policy-driven impacts oncircular economy of electric vehicle batteries in the United States. *Environment Systems and Decisions*, *45*, Article 4. https://doi.org/10.1007/s10669-024-09995-6SpringerLink
7. Zabetian-Hosseini, A., Joós, G., & Boulet, B. (2023). Control design for effective usage of electric vehicles in V2G-enabled DC charging stations. *IEEE Transactions on Power Delivery*, *38*(6), 4335–4346. https://doi.org/10.1109/TPWRD.2023.3245678SpringerLink
8. Juan, A. A., Mendez, C. A., Faulin, J., de Armas, J., & Grasman, S. E. (2016). Electric vehicles in logistics and transportation: A survey on emerging environmental, strategic, and operational challenges. *Energies*, *9*(2), 86. https://doi.org/10.3390/en9020086MDPI
9. Kashyap, A., Bajpai, A., & Upadhyay, R. K. (2023). Two-wheeler electric vehicles purchase policy. In *Transportation Systems Technology and Integrated Management* (pp. 399–442). Springer. https://doi.org/10.1007/978-981-99-1517-0_18SpringerLink
10. Charette, R. N. (2023). *The EV transition explained: Policy roadblocks*. IEEE Spectrum. https://spectrum.ieee.org/the-ev-transition-explained-658797681globalpolicy.ieee.org+2IEEE Spectrum+2IEEE Spectrum+2

Chapter 13

Safety and Standards

13.1 CRASH TESTS AND SAFETY STANDARDS IN ELECTRIC VEHICLES

Safety of electric vehicles (EVs) is one of the prime focus areas for regulators, makers, and general consumers. As a global phenomenon, EVs must now comply with stringent safety regulations to build public trust and prevent injury or fatalities in real-world scenarios. This chapter analyzes the principles and practices of crash testing and safety norms related to EVs. This technical overview provides an interpretation of testing protocols, differences from internal combustion engine vehicles (ICEVs), and a tabular compilation and database analysis.

Introduction to Crash Tests in EVs

Crash tests simulate accidents under controlled conditions. They evaluate the structural integrity of a vehicle in the event of a crash and its occupant protection capabilities. For EVs, these will also include damage or threats arising from battery punctures, electrical failures, and thermal runaway events.

There are several types of crash tests:

- Frontal impact tests
- Side impact tests
- Rear impact tests
- Rollover tests
- Pedestrian safety tests

Crash tests are carried out on the same environment for electric cars, but with more tests suited to be conducted with battery electric platforms.

Regulatory Bodies and Safety Standards

Regions have managed their specific vehicle safety protocols through their government agencies or independent organizations. Table 13.1

DOI: 10.1201/9781003608097-13

Table 13.1 Regulatory Bodies and Key Crash Safety Standards.

Regulatory Body	*Region*	*Standard Name*	*Applicable to EVs*	*Description*
National Highway Traffic Safety Administration (NHTSA)	United States	Federal Motor Vehicle Safety Standards (FMVSS)	Yes	Enforces crash tests, lighting, braking, airbags, and more
European New Car Assessment Programme (Euro NCAP)	Europe	NCAP Protocols	Yes	Provides star ratings based on occupant, child, and pedestrian protection
China New Car Assessment Program (C-NCAP)	China	GB 18384, GB/T 31467	Yes	Includes EV-specific thermal and electrical safety evaluations
Global Technical Regulations (GTR)	Global United Nations Economic Commission for Europe (UNECE WP.29)	GTR Nos. 13, 20, 21	Yes	Framework for international harmonization of safety standards
Bharat New Vehicle Safety Assessment Program (BNVSAP)	India	AIS-098, AIS-156	Yes	Includes crash tests and battery safety for Indian EVs

gives major regulatory bodies and applicable safety standards for EVs under crash tests.

Defined in those standards are specific crash test procedures, responses to be shown by the vehicle, acceptable intrusion levels, and occupant protection devices that must be provided. These are updated with time-concerning technological developments and new risk scenarios relevant to electric powertrains.

Particular Safety Concerns in EV Crash Testing

There are many similarities in crash testing procedures for both EVs and ICEVs; nonetheless, these have a few unique challenges.

Battery Damage and Fire Risk

A major issue with EV collisions is the integrity of the battery pack. High-energy lithium-ion batteries tend to thermal runaway after being crushed or otherwise deformed in the process of impact. Hence, crash tests incorporate instrumentation to monitor the temperature, potential leakage, and electrical insulation in post-crash settlings.

High-Voltage System Isolation

High-voltage systems in EVs must always be properly insulated and automatically disconnected in case of crash before any fault occurs. This is to reduce the possibility of electrocution of both occupants and first responders.

Structural Reinforcement

The presence of battery packs [1] in the floor of a vehicle usually results in the design of EVs with reinforced underbodies. Crash tests examine how well those structures distribute forces and are supposed to protect occupants, as well as batteries.

Types of Crash Tests and Their Objectives

Crash tests tend to reproduce certain real-world accident scenarios. An individual crash test gives different information about safety performance for the vehicle.

Frontal and side impact tests depicted in Table 13.2 will always remain the most significant from the perspective of safety ratings. However, with specific threats around batteries such as battery fire and high-voltage shock, rear impacts and rollover are now being treated with more scrutiny.

Table 13.2 Common Crash Test Types and Key Evaluation Metrics.

Test Type	*Description*	*Key Metrics Evaluated*	*EV-Specific Considerations*
Frontal offset	Vehicle hits a deformable barrier at 40–64 km/h	Chest injury, head injury, structural collapse	Battery deformation, cabin integrity
Side impact	Moving barrier hits the side at 50–60 km/h	Rib deflection, pelvis load, side curtain airbag performance	Lateral intrusion into battery area
Pole impact	Vehicle slides sideways into a fixed pole	Side torso impact, roof pillar performance	Structural failure near battery casing
Rear impact	Vehicle is hit from behind	Whiplash injuries, seat design, fire risk	Battery crush risk from rear deformation
Rollover	Vehicle flips onto its roof or side	Roof strength, restraint systems, ejection risk	Center of gravity impact due to battery placement

Crashworthiness of EVs Versus ICEVs

EVs differ from ICE cars in the structural layout, crash energy distribution, and component placement. Absence of a traditional engine allows crumple zone optimization, but the location of enormous batteries creates further challenges.

As shown in Table 13.3 although generally performing better in rollover tests, owing to their low center of gravity, the battery deformation management proves to be the challenge in laterally or rearward directed crashes. This creates the need for unique underbody engineering.

Safety Evaluation of Batteries Against Crash Scenarios

Crashing tests of EVs include, among different inspections after impact and simulations, those regarding battery system safety. These include the following.

Electrical Isolation Testing

It is ensured that there is no continuity between high-voltage circuits and the chassis of the vehicle after a crash.

Table 13.3 Structural Comparison—EVs Versus ICEVs.

Feature	*ICEVs*	*EVs*
Engine block	Present in the front	Absent
Battery pack	Small (12V) battery	Large lithium-ion pack in floor
Crumple zones	Front-biased	Often front and rear balanced
Center of gravity	Higher	Lower (due to floor-mounted battery)
Underbody reinforcement	Minimal	Heavily reinforced for battery protection

Thermal Stability Monitoring

A rise detected by post-crash temperature sensors indicates the beginning of thermal runaway.

Electrolyte Leakage Assessment

Leakage from damaged cells will be analyzed for flammability and chemical hazards.

As depicted in Table 13.4, many regulatory tests require even post-crash observation periods of as much as 24 hours in order to evaluate the possibility of late battery reactions. Safe designs prevent ignition events both immediately and slowly thereafter.

Global Crash Test Ratings for Leading EVs

Different international agencies have subjected various EV models to their tests. Table 13.5 summarizes the crash test ratings of popular EVs from different regions.

EVs nowadays are receiving top safety ratings, thereby proving that electric powertrains, with good designs, do not compromise occupant protections. However, the testing protocol varies, which might affect the total scores.

Future Trends in EV Crash Safety

Safety standards continue to evolve to accommodate changes in technology, as new risks and innovations evolve [2].

Solid-State Batteries

Compared with traditional lithium-ion batteries, they are more thermally stable, making them fire safe. Crash tests will require new protocols that follow those failure modes for crashes.

Table 13.4 Battery Safety Test Parameters After Crash.

Parameter	*Target Criteria*	*Test Instrument Used*	*Safety Thresholds (Typical)*
Voltage Isolation	≥ 500 Ohms/Volt	Megohmmeter	Isolation Resistance > 100 Ω/V
Temperature rise	≤ 5°C above ambient after 30 min	Thermal sensors	No rapid escalation
Electrolyte leakage	None visible or measurable	Visual + chemical analysis	No hazardous leakage allowed
Battery compartment integrity	No direct exposure	Structural inspection	No punctures or deformation

Table 13.5 Sample Crash Test Ratings of Popular EVs.

Model	*Euro NCAP Rating*	*NHTSA Rating (United States)*	*C-NCAP Rating (China)*	*Key Highlights*
Tesla Model 3	5 stars	5 stars	5 stars	Excellent structural performance, battery isolation
Hyundai Ioniq 5	5 stars	5 stars	4 stars	Strong side impact scores, slightly weaker rear crash
BYD Atto 3	5 stars	Not rated	5 stars	Strong cabin integrity, good battery protection
Nissan Leaf	5 stars	4 stars	4 stars	Adequate front crash, minor side pole concerns
Volkswagen ID.4	5 stars	5 stars	Not rated	Very low rollover risk, strong passenger cell

Vehicle-to-Grid Systems

With two-way flow of energy in vehicle-to-grid (V2G) systems, isolation and surge protection would become essential in post-crash situations.

New Advanced Materials

The use of lightweight composites and aluminum frames reduces weight but improves crash energy management to offset the weight of batteries.

AI-Based Crash Prediction

Some electric cars employ prediction algorithms to tighten seat belts or alter suspension just milliseconds before a crash [3].

Conclusion

Crash testing and standards for safety in EVs are one of the key pillars for the EV ecosystem. While normal cars take care of mechanical safety elements, from an EV perspective, the integration will require efforts toward electrical and chemical safety. Agencies worldwide have adjusted regulations to factor the risks pertaining to batteries, and manufacturers have responded with design enhancements and management of energy. The findings [4] from global crash test programs indicate that modern EVs have been able to meet and even exceed safety levels that were originally set for ICEVs. The safety features are expected to improve even further as technologies maturing in the processes like solid-state batteries and artificial intelligence (AI)-driven pre-collision systems.

Ensuring that EVs pass [5], besides conventional crash tests, any new battery safety and electrical isolation tests is critical for their acceptance by the public. While this scenario continues to develop, establishing crash-testing standards remains the highest priority in the safer and more sustainable mobility vision.

13.2 FIRE SAFETY AND ELECTRICAL HAZARDS IN ELECTRIC VEHICLES

EVs promise significant progresses in clean transportation, but they also pose new hazards regarding fire safety and high-voltage electrical systems. Because of the swelling size of lithium-ion battery packs and high-power electronics, EVs, unlike ICEVs, can lead to thermal runaway events or electrical faults under certain conditions. This section deals with fire safety and electrical hazards of EVs, their causes, safety measures, testing standards, and mitigation technologies.

Intro to Fire and Electrical Risks in EVs

EVs are statistically on [6] par with ICEVs regarding fire incidences; however, the nature of the fires and their methods of extinction differ significantly. Flammable electrolyte materials and high-voltage systems thus underline the need for specific protocols for risk prevention, detection, and response. Mechanical impact, battery overcharging, thermal runaway, or manufacturing defects may initiate fire incidents in EVs.

Major Causes of Fire and Electrical Hazards in EVs

Fire and electrical safety concerns are based on several key areas in the operation of EVs. These causes are often interrelated enough so that effective safety requires an integrated solution in monitoring and design.

Each cause listed in Table 13.6 contributes to either thermal risk or electrical safety risk. Understanding these pathways is crucial in designing layered protection systems and defining robust response protocols.

Battery Chemistry and Fire Behavior: the type of battery will impact the fire behavior response measures significantly. Lithium-ion batteries predominantly used in EVs have presumably very high energy densities but are also hazardous when burnt.

Lithium-Ion Battery Fire Characteristics

- Fires are called to be self-sustaining because of internal energy.
- The fire is capable of producing intense heat (over 1,000 degrees).
- Emission of toxic gases including hydrogen fluoride (HF).
- Reignition can occur hours after the initial extinguishment.

Emerging Alternatives

- Solid-state batteries have better thermal stability.
- Lithium–iron phosphate (LFP) batteries burn less easily.

NMC batteries as mentioned in Table 13.7 are commonly used but more thermally reactive, LFP batteries are heavy but much more stable and safer.

Table 13.6 Common Causes of Fire and Electrical Hazards in EVs.

Cause	*Description*	*Risk Factors*	*Typical Outcome*
Battery thermal runaway	Uncontrolled increase in battery temperature leading to fire	Cell defects, mechanical damage, overcharging	Fire, explosion, toxic gas release
Internal short circuits	Contact between internal battery electrodes	Dendrite growth, puncture, electrolyte leakage	Local heating, fire initiation
External electrical faults	Short circuits in power electronics or high-voltage wiring	Insulation failure, water ingress, improper installation	Electrocution risk, arc flash
Overcharging or overdischarging	Charging/discharging beyond safe voltage limits	Faulty battery management system (BMS), charger malfunction	Swelling, gas venting, fire
Crash-induced damage	Physical impact causes structural battery failure	High-speed collision, side impacts	Battery pack rupture, thermal runaway
Poor thermal management	Inadequate heat dissipation in battery packs	Hot climates, poor design, blocked cooling	Cell overheating, thermal imbalance

Table 13.7 Comparison of Battery Chemistries and Fire Characteristics.

Battery Type	*Energy Density (Wh/kg)*	*Thermal Stability*	*Fire Risk Level*	*Common Usage*
Lithium nickel manganese cobalt oxide (NMC)	200–250	Moderate (150–200°C)	High	Passenger cars, SUVs
LFP	140–160	High (>250°C)	Low	Buses, delivery vans
Lithium titanate (LTO)	70–80	Very high (>300°C)	Very low	Industrial, heavy-duty EVs
Solid-state (experimental)	250–300 (target)	Extremely high	Very low	Next-gen prototypes

Modifications to solid-state chemistries in point-cell production promise yet further reductions in fire hazard.

High-Voltage System Hazards

With respect to ICEVs rated at 12–48 volts, EVs operate electrical systems rated between 400 and 800 volts, thus posing new threats to passengers, mechanics, and first responders.

Electric Shock Risk

- Severe injury or death may occur through contact with damaged high-voltage lines.
- Isolation monitoring systems provide disconnection of power during faults.

Arc Flash and Short Circuiting

- High-energy electrical arcs can form during a fault, thus offering burns and electrical fire hazards.
- Good insulation and fuses are a must.

As shown in Table 13.8, High-voltage systems are thus subjected to multiple protective layers, including real-time diagnostics and automatic isolation capabilities, all for the minimization of risk in normal operation and in case of an impact.

Table 13.8 High-Voltage Electrical System Risk Factors and Controls.

Risk Factor	*Source*	*Mitigation Strategy*	*System Used*
Electric shock	Contact with high-voltage lines	Automatic disconnects, insulation monitoring	Battery disconnect units (BDU)
Arc flash	Sudden electrical discharge	Current-limiting fuses, arc shields	Fast-acting circuit breakers
Ground fault	Insulation failure	Ground fault detection systems	Isolation monitoring devices (IMDs)
Overcurrent/ short circuit	Faulty inverter or motor	Electronic protection circuits	Solid-state relays (SSRs), BMS control

Fire Testing Protocols for EV Batteries

More stringent testing protocols are designed to simulate fire and electrical failure conditions under controlled conditions by the regulatory statutory organizations. These tests help guarantee that the sections meet worldwide safety provisions prior to market authorization.

Thermal Abuse Tests: The batteries are subjected to extreme elevations in temperature to determine their capability of containing internal reactions.

Mechanical Abuse Tests: The crushing, penetrating, and vibrating conditions are simulated in order to assess structural integrity.

Electrical Abuse Tests: Charging, short-circuiting, and forced discharge tests used to evaluate battery safety and control systems [7].

These tests shown in Table 13.9 are important to validate the safety of battery packs in extreme conditions. Only after passing these evaluations are EVs cleared for commercial deployment in the respective jurisdictions.

Fire Detection and Suppression Systems in EVs

To mitigate fire hazards, today's EVs will have advanced fire protection, detection, and suppression systems.

Battery Management Systems (BMSs): They monitor cell voltages, temperatures, and currents. If any parameter exceeds safe thresholds, these systems can isolate or shut down affected modules.

Onboard Thermal Sensors: Thermistors and infrared (IR) sensors provide data on temperature in real time and allow for an early intervention.

Fire Suppression Techniques: Some EVs, especially commercial or fleet ones, integrate suppression agents (e.g., clean gas and foam) triggered to activate when fire is detected.

Table 13.9 EV Battery Fire Safety Test Protocols.

Test Type	*Description*	*Objective*	*Standard/Test Used*
Crush test	Apply force on battery cells	Detect physical deformation and rupture	UN 38.3, ECE R100 Annex 8
Nail penetration test	Insert metal nail into battery	Simulate internal short circuit	IEC 62660–2
Overcharge test	Charge battery beyond rated voltage	Test BMS response to electrical stress	SAE J2464, UL 2580
Fire exposure test	Expose battery to open flame	Assess flammability and containment	UN 38.3, GB/T 31467
Water immersion test	Submerge battery in water	Test electrical isolation under wet conditions	ISO 6469–1

Preventive systems like the BMS and fireproof materials substantially contribute to safety in the absence of an incident. Active suppression is more usually found in public transit and industrial EVs.

As shown in Table 13.10, even before the incident occurs, preventive systems and materials such as BMSs and fire-resistance materials play a significant role in assuring safety. Active suppression tends to be most commonly seen in public transport or industrial EVs during operation.

Safety Emergency Response and After Incident: The emergency response to any fire or electrical hazard that has arisen becomes critical at this point to limit the damage and obviate any secondary hazards.

Emergency Disconnects

High-voltage systems must include a manual cut-off or crash-activated isolator to improve safety for rescuers during EV incidents.

Firefighter Training

Training is being put into place internationally to assist first responders in the management of EV fires, thermal runaway, and high-voltage safety.

Reignition Risk

Batteries on fire can reignite hours or even days after complete cooling due to residual heat or internal reactions. Water application generally calls for a prolonged application procedure.

Table 13.10 EV Fire Prevention and Suppression Technologies.

Component	*Function*	*Technology Used*	*Common Application*
BMS	Monitors and controls cell behavior	Real-time diagnostics, fail-safes	All EVs
Thermal runaway detectors	Early warning of excessive heat in cells	Multipoint temperature sensors	High-capacity EVs
Battery enclosure systems	Protect against external fire or impact	Flame-retardant casing, pressure relief vents	Premium EVs, trucks
Suppression systems	Actively extinguish fire in case of detection	Aerosol, foam, inert gas systems	Buses, high-risk applications
Fire-resistant cabling	Prevent cable combustion and melting	Cross-linked polymers, shielded cables	Entire HV architecture

Global Regulatory Framework for Fire and Electrical Safety in EVs

Governments, together with standard bodies, have been active in refining the safety specifications for EVs currently under development. These frameworks serve purposes that include the following.

Cross-Border Compatibility

Notable Standards:

- UN ECE R100 Rev 2: Concerns uniform provisions about electric safety, including battery fire resistances.
- SAE J2929: Defines electrotechnical propulsion system safety of EVs and hybrid electric vehicles (HEVs).
- UL 2580: Fire safety requirements for battery in EVs currently under development.

Conclusion

Fire safety and electric hazard management in vehicles is perhaps one of the most under-considered discussions across chemistry, electrical engineering, mechanical design, and emergency response. Modern EVs are always outfitted with advanced systems that will detect, prevent, and suppress hazards most effectively. The introduction of high-voltage batteries does not mean that the risks are eliminated; they certainly require maximum innovation regulation and continuous training. Through battery chemistries, better fire detection, robust mechanical protections, and international safety standards, the bar continues to be raised for the EV industry in terms of fire and electrical safety. Adoption goes on; so, these safety guards are significant for functionality and for integrating EVs into society.

13.3 INDUSTRY STANDARDS AND REGULATIONS IN ELECTRIC VEHICLES

As EVs transition from niche to mainstream [8], safety regulations quickly became important alongside the need for standardization in technical protocols. These standards include measures ensuring safety, interoperability, environmental compliance, and consumer protection. Whereas internal combustion engine (ICE) cars have a uniform component mix, EVs have several new ones, including batteries, power electronics, and high-voltage cables and charging structures, and each of these areas requires its own set of technical regulations. This section [9] outlines the regulatory global framework of industry standards and policies governing EV safety and performance across different markets.

Standards and Regulations for EVs—Introduction

Industry standards and regulations for EVs are set by international, regional, and national bodies. These consist of a number of purposes:

- Establishment of baseline safety protocols for manufacturers and suppliers.
- Interoperability of parts and infrastructures (e.g., chargers and plugs).
- Facilitate certification and homologation of vehicles.
- Allow consumer trust through compliance and labeling.

Standards fall into three main categories:

- Vehicle safety/battery safety standards
- Charging infrastructure standards
- Regulatory compliance type approval and certifications

Global Standard-Setting Bodies for EVs

Different organizations participate in setting EV regulations. Some are compulsory, while others are considered voluntary benchmarks widely adopted by manufacturers.

As depicted in Table 13.11, these institutions are working together to lessen their overlapping technical barriers and inspire innovation, thus ensuring that uniform safety and operation benchmarks are met by EVs even across borders. While the UNECE frameworks mostly serve as an international base, they have to be regionally adapted to local necessities.

Safety Requirement for EVs

The safety of EVs includes both crash safety and electrical safety. It defines how acceptable systems will function in normal operations and during some incidents, such as accidents and fires.

Most Important EV Safety Standards

- ECE R100 (Rev.2): Concerns functional and electrical safety of battery systems for EVs.
- SAE J2929: Safety defined for propulsion battery systems.
- UL 2580: Tests whether or not the battery is safe under thermal, mechanical, and electrical abuse conditions.
- ISO 6469 Series: Electric safety (Part 1), functional safety (Part 2), and battery containment/fire (Part 3).

Table 13.11 Major Standard-Setting Bodies Relevant to EVs.

Organization	*Region/Scope*	*Focus Area*	*Examples of EV Standards/ Regulations*
UNECE WP.29	Global (United Nations)	Vehicle safety and environmental performance	ECE R100, R136, R85
International Electrotechnical Commission (IEC)	Global	Charging systems, battery testing	IEC 61851, IEC 62196, IEC 62660
Society of Automotive Engineers (SAE)	North America	Vehicle design, batteries, charging	SAE J1772, J2954, J2464
International Organization for Standardization (ISO)	Global	Functional safety, communication protocols	ISO 6469, ISO 15118
Bureau of Indian Standards (BIS)	India	National EV specifications	AIS-038, AIS-156, IS 17017
European Committee for Standardization (CEN)/European Committee for Electrotechnical Standardization (CENELEC)	Europe	Safety, electrical performance	EN 50604, EN 62133

As shown in Table 13.12, these standards work together in any process to ensure the operation of batteries under normal and fault conditions. Crash-triggered isolation systems, cell rupture thresholds, and fire propagation limits are all validated during compliance testing.

Charging Standards and Connector Regulations

Charging standards are indispensable [10] to ensure a safe and universal operation of EVs across varied infrastructure processes. This includes voltage compatibility, connector types, and communication between the vehicle and the charger.

Key Charging Standards

- IEC 61851: Provides definitions to the architecture of an electric charging system.
- IEC 62196: Plug and connector dimensions (Type 1, Type 2, CCS, CHAdeMO).
- SAE J1772: Standard connector for AC Level 1 and Level 2 charging in North America.
- ISO 15118: Smart charging and V2G system-enabled communication.

We see from Table 13.13, while the industry is tending to global convergence around the CCS standard and ISO 15118 for V2G applications,

Table 13.12 Selected Vehicle and Battery Safety Standards.

Standard	*Area of Focus*	*Description*	*Common Applications*
UNECE R100	Battery and high-voltage safety	Covers insulation, shock protection, fire resistance	Mandatory in many UN member states
SAE J2929	Battery safety	Mechanical, thermal, and chemical safety for lithium ions	North American EV certification
ISO 6469–3	Battery containment	Prevents leakage, thermal propagation, fire post-crash	Global OEM product development
UL 2580	Battery abuse and fire testing	Evaluates under impact, fire exposure, overcharge	US market access and insurance approval
AIS-156	Battery safety for Indian EVs	Includes thermal propagation and cell rupture protections	Required for sale in India

Table 13.13 Global EV Charging Standards and Connector Types.

Region	*Connector Type(s)*	*AC Standard*	*DC Standard*	*Communication Protocol*
North America	Type 1 (SAE J1772), CCS1	SAE J1772	CCS Combo 1	ISO 15118, PLC
Europe	Type 2, CCS2	IEC 62196–2	CCS Combo 2	ISO 15118, OCPP
China	GB/T	GB/T 20234	GB/T 27930	Proprietary/ISO 15118
Japan	Type 1, CHAdeMO	JARI	CHAdeMO	CAN protocol
India	Type 2, Bharat EV specs	Bharat AC001	Bharat DC001	IS 17017, ISO 15118

the regional legacy systems of CHAdeMO (Japan) and GB/T (China) still wield considerable influence. Vehicle type approval and certification regulations type approval is the procedure to get a vehicle model certified for sale in a particular market. The process covers crash tests, emissions compliance (hybrids), safety feature validation, and software audit. Regulatory Regimes by Region: Europe: UN/ECE type approval under Whole Vehicle Type Approval (WVTA); USA: self-certification by OEMs under FMVSS (NHTSA enforcement); India: certification by ARAI under Central Motor Vehicle Rules; China: certification by the Ministry of Industry and Information Technology (MIIT).

Regional type approval shown in Table 13.14, further works as a regulatory gatekeeping mechanism. It certifies that all EVs entering a market meet safety and environmental performance metrics required locally. Generally, the process involves lab tests, real-world simulations, and compliance documentations.

Cybersecurity and Functional Safety Standards: The present-day EVs are far more digitalized, since battery management, charging systems, and even some driver-assist features are software-driven. Therefore, cybersecurity and functional safety regulations have become paramount in the process. Key Functional and Cybersecurity Standards: ISO 26262: functional safety of road vehicles; ISO/SAE 21434: road vehicle cybersecurity engineering; UNECE WP.29 CSMS/R155 and SUMS/R156: mandate cybersecurity and software update frameworks for OEMs.

As seen from Table 13.15, the EVs connect to the grid, cloud services, and autonomous systems, functional and cybersecurity standards are crucial for the safety of the passenger and the protection of the infrastructure. These standards, in turn, cover risk classification, fail-safe design, and update tracking mechanisms. Harmonization of Global Standards: globalization of the automotive market makes harmonization of safety standards across the world a strategic imperative.

Table 13.14 Regional Type Approval Frameworks for EVs.

Region	*Approval Body*	*Legal Framework*	*Key Requirements*
European Union	National authorities (under ECE)	EU WVTA, ECE R standards	Battery safety, crash tests, EMC, noise
United States	NHTSA	FMVSS, EPA regulations	Crash testing, labeling, defect reporting
China	MIIT, CATARC	GB standards, local subsidies	Thermal safety, environmental tests, range checks
India	ARAI, ICAT	AIS, CMVR rules	Thermal runaway, charger compatibility
Japan	MLIT	JIS and JARI norms	Structural integrity, isolation, fire resistance

Table 13.15 Functional and Cybersecurity Regulations for EVs.

Standard/ Regulation	*Area Covered*	*Scope of Application*	*Mandatory Region(s)*
ISO 26262	Functional safety	Safety lifecycle for electrical/electronic systems	Global OEMs
ISO/SAE 21434	Cybersecurity in vehicle architecture	Risk assessment, software security	European Union, Japan, Korea
UNECE R155 (CSMS)	Cybersecurity management system	Governance, threat modeling	Europe (from 2024)
UNECE R156 (SUMS)	Software update management	Over-the-air (OTA) update security	Europe (from 2024)
AIS-189	Cybersecurity for Indian EVs	Aligned with ISO/ SAE 21434	India

Harmonized standards:

- Simpler procedures for vehicle export/import
- Lower manufacturing complexity and cost
- Globally accepted safety benchmarks

Some examples of harmonization initiatives include the following:

- The UNECE WP.29 framework promotes the mutual recognition of type approvals among 50+ countries.
- ISO and IEC are jointly developing global charging and battery standards.
- India's AIS standards are being progressively aligned with ECE/ISO protocols.

Full harmonization might still be years away, but a convergence trend is likely, particularly for functional safety, battery testing, and V2G communications.

Conclusion

The standards and regulations governing the industrial sector are the lifeblood of safety, efficiency, and reliability for universal EV ecosystems. Apart from safeguarding users and the infrastructure, they create an equitable playing field for the manufacturers by ensuring that technological advancements take place within the broader societal expectations of safety. The rapid pace of regulatory reforms on EVs is forcing the industry to address acute challenges in battery safety, high-voltage systems, charging interoperability, software updates, and cybersecurity. Adhering to these standards, automakers and suppliers are paving the way for a safer, stronger, and interconnected EV market on a global level. With the increasing push for harmonization, it is likely that greater convergence of standards will be witnessed in the future across regions, minimizing technical barriers to accelerate the adoption of electric mobility worldwide.

REFERENCES

1. Shi, L., Yan, X., & Huang, Z. (2025). Collision compatibility optimization of battery electric vehicles based on MPDB impact test. *International Journal of Automotive Technology*. https://doi.org/10.1007/s12239-025-00226-8
2. Sinha, N., Jain, V., Sehrawat, R., & Dhingra, S. (2025). Synergizing the future: Electric vehicles, artificial intelligence, and smart grids. *Smart Grids and Sustainable Energy*, 10(1). https://doi.org/10.1007/s40866-025-00247-3
3. Yang, S., Zhou, X., Zhou, S., Gao, Z., Cui, H., Hua, Y., Li, Q., & Chen, F. (2025). Advanced wireless battery management system for electric vehicles: Design, application and perspective. *Automotive Innovation*, 8(2), 421–442. https://doi.org/10.1007/s42154-024-00317-9
4. Aguado-Tevar, O., Diaz-Marcos, L., Delso-Vicente, A., & García de Blanes-Sebastián, M. (2025). Towards sustainable mobility: Factors influencing the use of electric vehicles in Spain. *Discover Sustainability*, 6(1). https://doi.org/10.1007/s43621-025-01117-9
5. Nguyen-Tien, V., Zhang, C., Strobl, E., & Elliott, R. J. R. (2025). The closing longevity gap between battery electric vehicles and internal combustion vehicles in Great Britain. *Nature Energy*, 10(3), 354–364. https://doi.org/10.1038/s41560-024-01698-1
6. Mohd Tohir, M. Z., & Martín-Gómez, C. (2025). Evaluating fire severity in electric vehicles and internal combustion engine vehicles: A statistical approach to heat release rates. *Fire Technology*. https://doi.org/10.1007/s10694-025-01711-3
7. Hwang, S., & Bae, S. (2024). Realistic drive cycle-based testing methodology for batteries in micro hybrid electric vehicles. *Journal of Electrical Engineering & Technology*, 20(3), 1383–1394. https://doi.org/10.1007/s42835-024-02101-5

8. Beloborodov, S. S., Gasho, E. G., & Nenashev, A. V. (2024). On the feasibility of transition to electric vehicles. *Power Technology and Engineering*, 58(1), 155–162. https://doi.org/10.1007/s10749-024-01789-9
9. Mondal, S., & Suman, N. K. (2025). Challenges in embracing electric vehicles: Exploring the policy landscape and way forward. *Transportation in Developing Economies*, 11(2). https://doi.org/10.1007/s40890-025-00248-y
10. Liu, Z., Xiang, X., & Feng, J. (2023). Tracing evolutionary trajectory of charging technologies in electric vehicles: Patent citation network analysis. *Environment, Development and Sustainability*, 26(5), 12789–12813. https://doi.org/10.1007/s10668-023-04011-4

Chapter 14

Consumer Education

14.1 CHOOSING THE RIGHT EV MODEL

Electric vehicle (EV) purchases include more than price and brand comparisons. EVs have several characteristics that determine user experience, total cost ownership (TCO), environmental impact, and long-term performance. This chapter will show private users and fleet managers how to choose an EV model based on their use habits, geography, technology, and budget.

Understand Personal or Operational Needs

An understanding of personal or operational requirements is perhaps the first and most fundamental step in selecting an EV model [1–3]. Assessing qualitative and quantitative needs is critical. Vehicle size, passenger count, average daily distance, and urban versus highway travel may affect personal charges [4,5]. Commercial customers and fleet operators need cargo capacity, uptime limits, charging, and operational range per charge. These constraints underpin all technical comparisons.

Compact battery electric vehicles (BEVs) with tiny batteries may be suitable for daily commutes under 60 km. For daily operations, a metropolitan distribution network logistics company may need electric vans or light-duty vehicles with over a 200-km range and fast-charging.

This approach reduces market-wide models to functionally relevant ones.

Classification and Availability of EV Types

Selecting an electric automobile requires knowledge of broad classes. BEVs, plug-in hybrid electric vehicles (PHEVs), hybrid electric vehicles (HEVs), and fuel cell electric vehicles (FCEVs) are the main EV types. Use cases have different infrastructure and operational impacts.

BEVs are fully powered by electric batteries, making them ideal for users with access to reliable charging. PHEVs blend an internal combustion engine (ICE) with an electric propulsion system, thus providing accommodation for users who are not yet fully ready to transit to EVs. HEVs charge

DOI: 10.1201/9781003608097-14

their batteries by regenerative braking and an ICE and provide very little range on electric-only power with no electrical recharging required. FCEVs use hydrogen fuel cells for power, which are still in their infancy in most stimulated markets, providing very limited refueling infrastructures.

A snapshot of some basic features about these EV types is provided in Table 14.1.

The choice of EV type, according to the table, should be made based on realistic considerations of local infrastructure and user pattern. Yes, a BEV would be the best choice for an intercity trip with a 400-km range, but you might want to consider a PHEV or an HEV if there are fast-charging stations scattered throughout the city! Plenty of high-utilization BEVs can then be accommodated in an urban center with robust EV policies and charging infrastructure.

Evaluating Battery Capacity and Range Requirements

Battery capacity measured, in kilowatt-hours (kWh), has direct opposing relations to the driving range offered by EVs. However, battery capacity, in isolation, is not a complete indicator for determining the appropriateness of a vehicle: it must be considered alongside efficiency, which is typically measured in kilowatt-hours per hundred kilometers (kWh/100 km). Costly trade-offs exist between the need for range, charging time, weight of the vehicle, and cost: large batteries are both expensive and take more time to charge. In this respect, Table 14.2 illustrates standard battery capacities and corresponding estimated ranges for different classes of EVs under routine operation.

A review of Table 14.2 shows that for urban driving, a compact hatchback could work; traveling more than 200 km to the countryside or intercity would be out of consideration. Also, cold weather and hilly terrain can cut range by 10–20%, which makes environmental context an essential consideration in the decision-making process.

Charging Infrastructure Compatibility

Compatibility to charging infrastructure is the third major selection criterion when judging an EV model. There are many forms of chargers differing in voltage, current, plug type, and speed of charging. Buyers need to validate the compatibility of the EV that they are purchasing with the charging infrastructure available; this includes home charging set-ups, public alternating current (AC) charging, and fast direct current (DC) charging stations. Table 14.3 summarizes the main EV charger types along with the average time per each type for fully charging an average battery.

This table shows that the degree of usefulness of a single EV model can change incredibly depending on the charging infrastructure. A Level

Table 14.1 Comparison of EV Types Based on Key Characteristics.

EV Type	*Primary Power Source*	*Charging Requirement*	*Electric-Only Range*	*Suitable Users*	*Infrastructure Dependency*
BEV	Battery pack	External charging	150–600 km	Urban commuters, long-distance drivers with access to charging stations	High (charging stations, home charging)
PHEV	Battery + ICE	Optional external charging	40–100 km	Suburban users with mixed travel needs	Moderate (charging helpful but not mandatory)
HEV	Battery + ICE	No external charging	2–5 km (electric only)	Users in transition, fuel economy conscious	Low (uses gasoline infrastructure)
FCEV	Hydrogen fuel cell	Hydrogen refueling	300–500 km	Early adopters in specific markets (e.g., California and Japan)	Very high (limited hydrogen stations)

Table 14.2 Battery Capacity Versus Range by EV Segment.

Vehicle Class	*Average Battery Size (kWh)*	*Estimated Range (km)*	*Typical Usage Scenario*
Compact hatchback	35–50	180–300	Urban commuting, short-distance travel
Mid-size sedan	60–80	300–450	Mixed city and highway usage
SUV/crossover	70–100	350–500	Family vehicle, intercity travel
Commercial van/truck	90–150	200–400	Last-mile delivery, utility services

Table 14.3 Charging Types and Time Estimates.

Charging Type	*Voltage/ Current*	*Charging Power*	*Time to Full Charge (60 kWh Battery)*	*Common Locations*
Level 1 (AC)	120V/12A	1.4 kW	35–50 hours	Residential (standard outlets)
Level 2 (AC)	240V/30A	7.2 kW	8–10 hours	Homes, work-places, malls
Level 3 (DC fast)	400–900V/ 100–500A	50–350 kW	15–60 minutes	Highways, ser-vice stations

2 charger should facilitate weekend charging for users based in the city, whereas for intercity users, fast DC chargers would reduce time on the road during work. The buyer should also check the charging connector types (CCS, CHAdeMO, Type 2, etc.) to avoid complications during work.

Cost Analysis: Purchase Price, Incentives, and Operational Costs

Although many people consider EVs to be pricier than internal combustion engine vehicles (ICEVs), in many cases, the initial purchase price has remained higher for many EVs. However, this is countered with subsidies, tax rebates, and deductions from registration in various jurisdictions. A sample analysis of comparison of costs of EVs and ICEVs for five years under real-world driving conditions is shown in Table 14.4.

Cost comparisons in Table 14.4 are significant for assessing TCOs over the life of the vehicle. The initial purchase price of an EV may seem high, but lower fuel and maintenance costs, in addition to subsidies, often bring the TCO in EVs below that of combustion engine vehicles.

Brand Reliability, Warranty, and After-Sales Support

The choice of an EV model must be influenced by brand reputation, warranty coverage of the battery pack, and the level and quality of the after-sales support offered. Battery degradation remains an issue for long-time owners despite much progress in the recent battery chemistries. Therefore, batteries are quite often guaranteed up to 8–10 years and/or 160,000 km by reputable manufacturers, assuring buyers about battery performance. Also, it pays to check out the network of service shops available, responsiveness of customer care, and extent of cover under the extended service programs. Table 14.5 compares the warranties provided by selected global EV manufacturers and illustrates the variations in consumer protection among them.

Table 14.4 Comparative Total Cost of Ownership over 5 Years.

Cost Component	*Mid-Size ICEV*	*Mid-Size BEV*
Purchase price	$30,000	$38,000
Government rebate	$0	−$5,000
Fuel/energy cost	$9,000	$2,000
Maintenance cost	$4,500	$1,800
Insurance premium	$5,000	$5,500
Total TCO	**$48,500**	**$42,300**

Table 14.5 Battery Warranty Comparison across Selected Manufacturers.

Manufacturer	*Battery Warranty (Years)*	*Distance Coverage (km)*	*Degradation Clause*
Tesla	8	160,000	Minimum 70% capacity retention
Hyundai	8	160,000	Minimum 70% capacity retention
Nissan	8	160,000	Minimum 70% capacity retention
BMW	8	100,000	Minimum 70% capacity retention

For the most part, warranty provision does not vary much among the main brands, but the availability of certified service centers does. People located in far-flung or underserved areas might opt for brands that have a larger coverage network. Selecting the right EV model demands multi-criteria examination beyond car purchasing [6–9]. A systematic assessment of personal or corporate mobility demands, electric car classes, battery range variances, charging compatibility, cost components, manufacturer reliability, and warranty can aid future-proof decision-making. Switching to electric mobility requires procedure, behavior, and infrastructural adjustments. Proper model selection would maximize EV economy, performance, and long-term economic and environmental benefits from this move. EVs are becoming popular in personal and business transport ecosystems; therefore, consumer education must include charging economics. Fuel rates are usually fixed per liter or gallon at the pump; however, EV charging costs vary owing to technical, geographic, temporal, and infrastructural factors. This chapter aims to provide readers with a comprehensive framework on the economic structure of EV charging costs, fuel cost comparisons, and how various charging contexts—from residential system set-ups to public fast charging—bleed into the economics of owning and operating an EV.

14.2 UNDERSTANDING CHARGING COSTS

Components of EV Charging Costs

Normally, the cost of EV charging is computed based on the amount of electricity consumed in a charging session and is represented in kWh. Nevertheless, the actual price that the consumer pays is affected by various factors: base electricity rate, time-of-use pricing structures, added services or network charges, charging speed levels, and demand charges at times, in some commercial stations. Residential users usually pay the rate for electricity that applies to their household usage, although it can actually vary depending on the time-of-use program or tiered energy slabs. Public charging stations often operate on a premium, not only to cover the cost of electricity but also factoring in the other overhead costs of infrastructure maintenance, transaction handling, and availability of service. In some parts of the world, there is an additional idle fee for users if the charger remains connected to the vehicle that is fully charged. To enhance the understanding of the relationship between these components, Table 14.6 provides a typical breakdown of the cost elements involved in residential and public charging.

From Table 14.6, it is evident that home charging tends to be the most cost-effective and stable, particularly if the users have fixed-rate electricity tariff plans [10,11]. On the opposite end of the spectrum, public fast-charging stations will have more dynamic pricing mechanisms encompassing fixed service charges and charges for exceeding the allotted charging window. The costs attributable to the vast installation and maintenance of high-capacity fast chargers are thus reflected in the high rates.

Home Charging Versus Public Charging Economics

For most users, home charging remains the most economical and convenient option. This method permits full overnight charging, ensuring the vehicle is completely charged and available each morning. Home charging costs vary primarily depending on local electricity rates, which could be on a flat, tiered usage slab, or a time-of-use basis. For time-of-use users, these savings stem from significant reductions when charging during off-peak rates. On the other hand, while costs on public charging may be required during extended trips or for urban dwellers without dedicated parking spaces, these are usually very high in kWh rates. For clarity on this difference, a comparative study of the average monthly cost of charging a mid-size EV over 1,200 km/month is presented in Table 14.7, which harmonizes different charging arrangements.

Table 14.7 presents clear evidence that the greatest savings accrue to those charging at home. Those who mainly rely on fast public charging may incur costs close to four times those charged at home under off-peak residential rates. Such public stations are important to guarantee flexibility

Table 14.6 Components of EV Charging Costs in Different Scenarios.

Charging Location	*Base Electricity Rate (USD/kWh)*	*Time-of-Use Variability*	*Additional Fees*	*Idle/Overstay Penalty*	*Infrastructure Impact*
Home charging	0.10–0.20	Yes	None	Rare	Low
Workplace	0.00–0.15 (subsidized)	Occasionally	Nominal	Rare	Moderate
Public Level 2	0.20–0.40	Yes	$0.50–$2/session	Sometimes	Moderate
DC fast charging	0.30–0.60	Yes	$1–$5/session	Common	High

Table 14.7 Average Monthly Charging Costs Based on the Charging Mode.

Charging Mode	*Electricity Rate (USD/kWh)*	*Efficiency (kWh/100 km)*	*Monthly Distance (km)*	*Monthly Energy Consumption (kWh)*	*Estimated Monthly Cost (USD)*
Home charging (flat rate)	0.13	15	1,200	180	23.40
Home charging (TOU off-peak)	0.09	15	1,200	180	16.20
Public Level 2 charging	0.30	15	1,200	180	54.00
DC fast charging	0.50	15	1,200	180	90.00

and mobility, but higher prices make them more challenging in long-term strategic considerations of EV ownership, especially for high mileage users.

Impact of Regional Electricity Tariffs

The electric tariffs vary tremendously, not only between countries but also within the same country across states, provinces, and cities. In a deregulated market, utility providers have the liberty to attach diverse types of plans to their business, such as flat-rate, peak-time surcharges, and dynamic pricing criteria tied to grid demand. Thus, in one location, the cost of charging an EV can be much different from another, even with the same vehicle and same charging gadgets. Table 14.8 demonstrates how electricity rates differ from region to region in some selected countries to show the worldwide variation in the affordability of EV charging.

Table 14.8 shows that the costs of charging an EV are local price-dependent variations in the electricity price. This means the operating cost advantage of the EVs over the ICEVs is finally nullified in countries like Germany and Japan, where higher costs of electricity are incurred. On the other hand, low electricity prices found in markets like India and Norway make EVs increasingly more cost-effective and affordable in the long term.

Effect of the Efficient Vehicle on the Costs of Charging

An important consideration directly determining the price of charging is energy efficiency. Different EV models consume different amounts of energy within kilometers due to factors such as size, aerodynamics, motor design, and battery management systems. Smaller hatchbacks or sedans are usually more energy-efficient compared to larger SUVs and performance-driving vehicles. To show the relationship between the types of vehicles and energy efficiency, Table 14.9 gives the average energy efficiency values for popular classes of EVs.

Table 14.8 Average Residential Electricity Rates in Selected Countries (USD/kWh).

Country	*Average Residential Rate ($)*	*Typical Monthly EV Charging Cost ($, 1,200 km @ 15 kWh/100 km)*
United States	0.15	27.00
Germany	0.38	68.40
Norway	0.12	21.60
India	0.08	14.40
Japan	0.26	46.80

Table 14.9 Typical Energy Efficiency by Vehicle Class.

Vehicle Type	*Average Efficiency (kWh/100 km)*	*Monthly Energy Use (1,200 km)*	*Cost @ USD 0.15/kWh*
Compact EV hatchback	13	156	23.40
Mid-size sedan	15	180	27.00
Crossover SUV	17	204	30.60
Full-size SUV/truck	21	252	37.80

As indicated in Table 14.9, the selection of an EV model leads to significant differences in monthly charging costs, even when electricity prices remain constant. The consumer whose full-size electric SUV is in operation will have to pay a 60% higher electricity bill than a compact EV driver because of the differences in efficiencies of the motor vehicles. These costs become even more substantial over multiyear ownership periods, which reinforces the importance of vehicle selection aligning with usage needs.

Smart Charging, Time-of-Use, and Demand Management

EV charging costs may also be reduced by using smart charging solutions. The provision of smart chargers allows users to arrange their charging sessions to coincide with off-peak hours, which offer lower electricity rates—usually linked with the time-of-use billing system adopted by their utility provider. Other regions have demand–response programs that give monetary incentives to users who shift their charging patterns in accordance with grid capacity. Table 14.10 provides a comparison of the savings from smart charging and static and time-optimized charging behavior for a typical residential user.

Table 14.10 shows savings resulting from the integration of smart charging into the daily routine. Even a user who schedules his charging automatically to take advantage of off-peak prices could save more than half the costs incurred in charging his EV. Demand–response programs, where available, often contribute to even more significant savings, realizing such benefits for both the individual consumer and the grid. Charging cost constitutes one of the most significant yet misunderstood facets in a broader discussion of EV economics. Unlike the fixed and pre-assured gasoline prices, the expenses for charging an EV vary tremendously according to many situational but dynamic factors. Electricity prices, vehicle efficiency, infrastructure, and customer behavior change. These choices assist users maximize EV investment value. Home charging is the cheapest, especially with a time-of-use electricity plan or smart charging system, but public charging stations are needed for long-distance travel and urban EV adoption. Strategic use is needed due

Table 14.10 Cost Comparison of Static Versus Smart Charging.

Charging Strategy	*Off-Peak Rate (USD/kWh)*	*Peak Rate (USD/kWh)*	*Charging Window*	*Estimated Monthly Cost*
Unscheduled charging	N/A	0.20	Any time	36.00
Smart scheduled charging	0.09	0.20	10 PM–6 AM	16.20
Demand–response enabled	0.06 (incentivized)	0.20	Utility-directed	10.80

to high unit costs and levies. Vehicle type and energy efficiency affect charging prices: buy accordingly.

14.3 MANAGING RANGE ANXIETY

Range anxiety is the most discussed psychological and practical barrier to EV adoption. Drivers experience range anxiety when they fear their EV may run out of power before they reach their destination or a charging outlet. ICEVs have a dense, mature gas station network, whereas EVs have infrastructure, charging time, and range forecast difficulties. Inexperienced EV drivers or consumers unfamiliar with the vehicle's real-world capacity may sense limited mobility freedom due to range anxiety.

Range anxiety is caused by battery performance, location, weather, travel planning, and charging infrastructure. Consumers will need technology and information to control fear. Range anxiety's causes, effects on driver behavior and adoption, and treatments are covered in this section. Understanding and controlling range anxiety may help EV owners feel more confident and integrate into mainstream transportation.

Origins and Psychological Basis of Range Anxiety

Range concerns are psychologically caused by humans exaggerating risk in uncertainty. While fuel-based transportation is something many people have lived with for decades, battery-based movement is a recent paradigm and not one most drivers are familiar with. This may be the primary consideration as gas stations are pretty much everywhere, and refills of traditional vehicles happen in no time. Given the lengthy deliberations on travel plans EVs impose, it is perhaps the drivers' perceived diminishing control of such journeys that adds stress. In most cases, this perception of stress is worsened by ignorance of the range indicators for the particular car, inconsistencies in automotive conditions, and the simultaneous lack of visibility and reliability of public charging stations.

Range anxiety can also be influenced by personal habits, geographic location, and lifestyle factors. For example, a driver who resides in a rural area with limited charging infrastructure would likely feel more anxiety than a city dweller who has access to many charging points. Likewise, anybody who travels long distances regularly or snaps at their schedule might have more difficulty accepting the limitations of EV range due to more personal conflict. It is by understanding these user-specific triggering instances that viable solutions can be engineered.

Real-World Driving Range Factors

The rated range provided by manufacturers usually pertains to the optimal conditions reached under standardized testing protocols. However, actual driving ranges may vastly differ, owing to various environmental and operational factors such as temperature, terrain, speed, payload, and auxiliary energy consumption. For example, high-speed highway driving dissipates more energy than urban stop–start driving due to aerodynamic drag and lesser opportunities for regenerative braking. Likewise, extreme temperatures would require either heating or cooling of the cabin, both of which draw on battery power, thereby reducing the available driving range.

In this respect, to illustrate the impact of differing factors on range, Table 14.11 presents a comparative analysis of estimated reductions in range for different conditions of a mid-size EV that achieves a nominal range of 400 km.

Table 14.11 shows that real-world conditions can diminish more than 10–25% effective EV range. This uncertainty is a significant source of anxiety and becomes a major contributor to the driver's inability to estimate remaining charge and confidently plan trips. Therefore, user education about the interaction of these factors with vehicle operation is an important part of any range anxiety management program.

Table 14.11 Estimated Range Reductions under Variable Driving Conditions.

Driving Condition	*Range Reduction (%)*	*Adjusted Range (km)*
Highway driving @ 120 km/h	20	320
Cold weather (<−5°C)	25	300
Hot weather with A/C use	15	340
Full payload (5 passengers + cargo)	10	360
Hilly terrain (frequent elevation)	18	328

Advances in Navigation and Energy Prediction Systems

Modern EVs are increasingly equipped with sophisticated energy prediction algorithms and navigation aids that dynamically adjust range estimates based on driving behavior, elevation changes, weather conditions, and traffic patterns. The real-time input will suggest routes that balance energy use and charging station access. Full trip planners can forecast wait times by recommending charging locations based on station occupancy and queue length.

Range anxiety necessitates this advanced talent, which devolves test intelligence. Dynamic prediction systems use real-time data to provide a more accurate and dependable vehicle performance estimate than static range displays. These devices can pre-plan routes and track battery life with smartphone apps.

The functionality and user benefits of static range indicators and dynamic energy prediction systems are presented in Table 14.12.

The table illustrates the counteraction of range anxiety by predictive navigation systems. Such technologies replace cautious or outmoded range estimations with context-sensitive, real-time information, giving drivers more self-control and confidence. To boost EV adoption and user experience, automakers and software companies invest extensively in this industry.

Density and Accessibility of Charging Infrastructure

Range anxiety is reduced by charging infrastructure density and accessibility. With high densities of chargers, maintained well at all times, transparent pricing, and smooth payment solutions, uncertainty is significantly reduced and consumer confidence rests assured. Mixed speed infrastructure that includes Level 2 chargers for everyday topping-off and DC fast chargers for time-sensitive travel allows for routing flexibility and short planning horizons in the process. Sparse or unreliable networks could, however, serve to heighten anxiety regarding the drivers, especially in unfamiliar places and rural settings. Table 14.13 lists ratios of chargers to vehicles in selected

Table 14.12 Static Range Indicators Versus Dynamic Energy Prediction Systems.

Feature	*Static Indicators*	*Dynamic Prediction Systems*
Adjusts for driving behavior	No	Yes
Considers terrain and weather	No	Yes
Real-time route optimization	No	Yes
Charging station integration	Limited	Comprehensive
Predictive accuracy	Low	High

countries in light of infrastructure density and self-reported levels of range anxiety.

The data captured in Table 14.13 reveals the inverse correlation between the availability of infrastructure and the level of anxiety in consumers. Countries with dense and reliable charging infrastructures report very low levels of concern and anxiety with respect to range. As the world embraces EVs, investment in scalable infrastructure that will be highly available for public use will go a long way in boosting consumer confidence and reducing any perceived risks.

Behavioral and Strategic Charging Practices

User behavior is an important aspect in managing range anxiety. Educating drivers to adopt effective charging behaviors—such as regular scheduled home charging, keeping battery levels above 10%, and opportunistic charging when doing errands—would reduce range anxiety. As users become familiar with their vehicles' real-world performance, they become more familiar with, and sense, the reliability of their systems; hence, there is a significant decrease in range anxiety due to experience.

Strategic trip planning is also a very good tool. Route planning including known charger locations, preferred speeds, and buffer margins for energy reserves would make the trip efficient and reduce risk. A driver could, for example, decide to stop at a fast charger when the charge goes below 30% instead of below 10% so that the flexibility is retained for some unavoidable detours or station unavailability.

Table 14.14 presents the practical effects of the adoption of charging behavior by comparing two strategies on average state-of-charge (SoC) safety margins and reported stress levels.

Table 14.14 shows that an individual adopting a cautious charging strategy is likely to keep the consumption level higher above the safety margin and generally feel less stress. Although this approach might demand more frequent but shorter charging sessions, it assures one's conscience and overall satisfaction with the experiences of EV ownerships.

Table 14.13 Charger-to-EV Ratios and Range Anxiety Trends.

Country	*EVs per Public Charger*	*Fast Chargers per 100 km of Highway*	*Reported Range Anxiety (Survey %)*
Netherlands	5.2	4.5	18
Norway	6.8	3.8	20
United States	18.4	1.2	47
India	35.0	0.6	53
Japan	7.1	3.0	22

Table 14.14 Comparison of Conservative Versus Risk-Tolerant Charging Behavior.

Charging Strategy	*Minimum SoC before Charging (%)*	*Average SoC at Charger Arrival (%)*	*User-Reported Stress Level (1–5)*
Conservative charging	30	45	1.8
Risk-tolerant charging	10	15	4.1

Role of Vehicle Range Extension Technology

New advances in battery technology and vehicle energy management are still improving the performance of EVs and reducing the incidence of range anxiety. More energy-dense batteries, more efficient powertrains, and better thermal management systems extend range and improve performance in adverse conditions. Some models provide additional range-extending features, such as solar roof panels or small onboard generators. Some of these features have not yet moved beyond niche applications.

Beyond hardware updates, software updates improve vehicle efficiency and constitute additional capabilities for range management. Over-the-air (OTA) updates may optimize battery consumption, refine predictive algorithms, or change regenerative brake settings to improve the range while avoiding physical alterations. The continuous technological improvements contribute to more reliability and trust in the vehicle range estimate. Table 14.15 outlines industry advancement of EVs regarding driving range across vehicle generations. The statistics presented in Table 14.15 shows that almost all earlier EV ranges within a decade tend to be improving in terms of battery capacity and thus range. With this progressive outlook, range anxiety will be very rare, especially for users driving less than 300 km a day, which constitutes the majority cases of daily usages.

While purchasing an EV is a radical departure from conventional vehicle purchase norms, refueling, maintenance, and performance expectations for ICEVs have all been very well established. EVs use a platform for the integration of battery chemistry with power electronics, software-driven energy management, and charging infrastructure.

14.4 TIPS FOR ELECTRIC VEHICLE BUYERS

Without proper translation, this new environment may have a steep learning curve for future buyers, resulting in incorrect decisions and ownership unhappiness. Technical evaluation, charging setup sizing, cost forecasts, legislative incentives, and long-term usability for prospective EV consumers will be covered in this section. Buyers can make future-proof purchases.

Table 14.15 Evolution of EV Ranges over Time (Select Models).

Model Name	*Year of Release*	*Battery Size (kWh)*	*Official Range (km, worldwide harmonized light vehicles test procedure [WLTP])*
Nissan Leaf (Gen 1)	2011	24	117
BMW i3 (Facelift)	2016	33	183
Hyundai Kona EV	2019	64	449
Tesla Model Y LR	2021	75	533
Lucid Air Touring	2023	92	620

Analyzing Driving Needs for Range Suitability

A decent driving evaluation should include EV purchase driving criteria. This method creates a realistic driving behavior profile to match vehicle range with real-world use cases by examining daily commute distance, weekly travel patterns, scheduled road trips, climatic conditions, and road infrastructure demands. Most modern EVs can drive 300–500 km per charge, depending on the topography, climate, and driving habits. For example, choosing a vehicle with too much range might cause unnecessary expenditure, while a vehicle with insufficient range could result in operational anxiety for the driver. In order to facilitate this matching, Table 14.16 presents a variable mapping of typical driving needs and inappropriate EV range categories.

Buyers should refrain from simply purchasing the highest-range vehicle available, as large battery packs come with higher costs, longer charging times, and greater environmental impact due to increased raw material usage. Instead, the emphasis should be on achieving an EV with a range that satisfies nominal daily requirements, including a safety buffer, while avoiding excessive energy storage capacity.

Battery Chemistry, Degradation, and Warranty Considerations

The battery pack is one of the most crucial and most expensive components in an EV, and this dictates not only the vehicle's range but also the lifetime, performance, and resale value. Different manufacturers use different battery chemistries: lithium nickel manganese cobalt oxide (NMC), lithium iron phosphate (LFP), or solid-state battery prototypes. There are differences between chemistries with regard to energy density, thermal stability, charge–discharge cycle life, and environmental effects. The capacity decreases with time due to battery degradation caused by usage patterns, fast-charging, ambient temperatures, and the depth of discharge. A thorough examination

Table 14.16 Matching Driving Profiles with EV Range Categories.

Driving Profile	*Average Daily Distance (km)*	*Recommended Range (km)*	*Suitable Vehicle Segment*
Urban commuter	20–60	150–250	Compact hatchback, mini EV
Suburban/regional driver	60–120	250–400	Mid-size sedan, crossover SUV
Long-distance traveler	120–250	400–600	Full-size SUV, premium EV sedan
Commercial fleet operations	80–200 (daily load cycles)	300–500	Electric vans, utility trucks

of manufacturer warranty and degradation policy should be done by a potential buyer. Generally, most major manufacturers offer an 8-year or 160,000-kilometer warranty, usually with some promise of retaining 70% battery capacity by the end of the term. The terms and conditions enlisted may greatly affect the long-term utility of the vehicle. Table 14.17 presents the comparison of the basic characteristics of different chemistries most commonly used in EV batteries.

As noted from Table 14.17, the buyer has to keep in mind usage expectations when selecting batteries. For instance, urban users with particular end use charging conditions and short daily range may benefit from vehicles based on LFP chemistry because of the long life style and the thermal robustness associated with it, while long-distance range users concerned with energy density will opt for NMC systems.

Charging Setup: Home Installation and Public Access App

The charging ecosystem is one of the most vital but relatively underrated aspects when it comes to owning an EV. Buyers should check whether their house supports either Level 1 (standard wall socket) or Level 2 (dedicated home charger) installations. With regard to Level 1 charging, no upfront investment for any infrastructure is necessary, and it offers relatively low charging power, typically giving only 6–10 km of range per hour, which is hardly useful for heavy users. Level 2 charges run on a 240V circuit and offer an extremely fast replenishment rate of around 25–50 km/hour, allowing complete charging of the battery overnight. Installation may involve electrical upgrades, permits, and relation to utility providers.

Besides home installations, buyers should also consider public charging sites relative to their coverage, availability, reliability, and accessibility within a given region. A big concern, especially for those who depend on intercity travel, would be compatibility with the fast-charging network and

Table 14.17 Comparison of Common EV Battery Chemistries.

Battery Chemistry	*Energy Density (Wh/kg)*	*Cycle Life (Full Cycles)*	*Thermal Stability*	*Fast Charging Suitability*	*Notable Users*
NMC	180–250	1,000–1,500	Moderate	High	Tesla (early models), Kia, Hyundai
LFP	120–160	2,500–3,000	High	Moderate	Tesla (China), BYD, MG
Solid-state (R&D)	>300	>3,000	Very high	Expected high	Toyota (pilot), QuantumScape

connectors: CCS or CHAdeMO. Mobile apps showing the real-time status of chargers, pricing, and possible reservations of chargers can greatly enhance user experiences. Table 14.18 summarizes the important distinctions between typical residential and public charging structures for the right infrastructure decisions.

Table 14.18 shows that charging strategy highly depends on individual driving habits and type of habitation. The apartment dweller or renter who does not have access to a home charger would depend on workplace or public charging stations; then public infrastructure becomes more vital in deciding factors in the purchase.

Total Cost of Ownership and Incentive Awareness

Perhaps among the most promising advantages of EVs is their very low total cost of ownership compared to ICEVs. Some EV models still have higher purchase prices initially, but some government subsidies, lower maintenance, and reduced fuel costs often make them cheaper in a multicompany or multiyear horizon. The buyer should consider the purchase incentives, tax rebates, reduced registration fees, and preferential access to restricted driving zones [12,13] or toll-free roads for a complete financial picture. In addition, buyers should consider that while EV insurance premiums may be higher due to increased repair costs, this is often offset by fewer mechanical breakdowns and no engine servicing. Fleet operators or business users may also benefit from accelerated depreciation schedules and fleet electrification grants. Table 14.19 presents a hypothetical comparison of TCO over five years between a mid-sized EV and similar ICE models. This results from assessments made in the preceding table.

The analysis in Table 14.19 shows that the initial higher investment in an EV is more than offset by lower running costs and incentives throughout the vehicle's life. Buyers should do TCO calculations tailored to their specific situations on local electricity rates, fuel costs, and any applicable policies before proceeding with financial decisions.

Table 14.18 Comparison of EV Charging Options.

Charging Type	*Voltage/Power*	*Installation Cost (USD)*	*Charging Speed (km/hour)*	*Ideal Use Case*
Level 1 (AC)	120V/1.4 kW	None	6–10	Occasional use, low-mileage users
Level 2 (AC)	240V/7–22 kW	500–1,500	25–50	Daily home charging
DC fast charging	400–800V/50–350 kW	None (public infrastructure)	150–800+	Road trips, commercial operations

Table 14.19 Comparison of the Five-Year Total Cost of Ownership.

Cost Component	*ICEV (USD)*	*EV (USD)*
Purchase price	30,000	38,000
Government incentive	0	–5,000
Fuel/energy cost	9,000	2,200
Maintenance and repairs	5,000	1,800
Insurance (5 years)	5,500	6,000
Resale value (after 5 years)	9,000	13,000
Net 5-Year TCO	**40,500**	**33,000**

Software Functions, Connectivity, and After-Sales Ecosystem

The major difference between electric cars today is the software architecture and services associated with connecting them. Buyers should keep an eye on the vehicle's user interface, updated policies, and mobile app integration to monitor charging, schedule, and perform remote diagnostics. OTA update, regenerative braking control, adaptive navigation according to charge levels, and vehicle-to-grid (V2G) readiness not only complement the use but also increase resale appeal.

Prospective owners should also evaluate after-sales ecosystems, which include the availability of service centers, battery replacement or repair procedures, digital service records, and mobile support options. As a rule, the brands that have older EV portfolios have better-trained technicians, speedier availability of spare parts, and broader service coverage. Table 14.20 summarizes some selected software and support features that should be considered by buyers during the evaluation process.

Table 14.20 shows that software and after-sales support features are no longer ancillary but core for today's EV experience. Buyers investing time understanding these systems would be better positioned to maximize their vehicles' potential across the lifespan of operations.

Table 14.20 Key Software and After-Sales Features in EV Ownership.

Feature	*Description and Benefit*	*Relevance to Buyer*
OTA updates	Remote software upgrades for performance and features	Extends vehicle lifespan
Mobile app integration	Monitoring, remote climate control, charging management	Enhances usability
Regenerative braking settings	Adjustable energy recovery during deceleration	Improves driving efficiency
V2G compatibility	Enables energy export to grid or home	Supports energy savings
EV-specific service network	Dedicated technicians and tools for EV diagnostics	Ensures maintenance quality

Transitioning to electric mobility requires becoming more nuanced in buying EVs than conventional vehicles [14,15]. Buyers should evaluate daily driving patterns, automobile range, battery chemistry and degradation, charging infrastructure, and long-term costs. Beyond mechanical specs, software, networking, and ecosystem maturity affect decision-making.

Data and contextual understanding of personal or fleet needs help satisfy EV buyers and increase profitability. As EV technologies and infrastructure improve, the process should be simplified and standardized. Customers require extensive knowledge to make healthy early and mid-adopter choices that support transportation sustainability till then [16,17].

REFERENCES

1. Ghali, Z. (2023). Determinants of ethically minded consumer behavior in relation to climate change: A cognitive hierarchal model perspective. *Environment, Development and Sustainability*, 27(3), 6225–6253. https://doi.org/10.1007/s10668-023-04128-6
2. Mustafa, S., Shi, Y., Adan, D. E., Luo, W., & Al Humdan, E. (2024). Role of environmental awareness & self-identification expressiveness in electric-vehicle adoption. *Transportation*. https://doi.org/10.1007/s11116-024-10515-3
3. Ngoc, A. M., & Nishiuchi, H. (2023). Joint effects of perceived benefits, perceived barriers, and policy interventions on behavioral intentions toward electric vehicles in Vietnam. *International Journal of Intelligent Transportation Systems Research*, 22(1), 94–107. https://doi.org/10.1007/s13177-023-00380-2
4. Zhao, X., Li, X., Wu, Y., Qiao, L., & Zhang, C. (2023). Assessment of the effects of China's new energy vehicle industry policies: From the perspective of moderating effect of consumer characteristics. *Environment, Development and Sustainability*, 27(2), 4319–4340. https://doi.org/10.1007/s10668-023-04078-z

5. Shbool, M. A., Alhmsi, K., Amr, M., Hajeer, R., Hamed, A., Al-Bazi, A., & Rawabdeh, M. (2025). Modeling consumer behavior and forecasting the automobile market: A system dynamics approach for sustainable mobility. *Arabian Journal for Science and Engineering*. https://doi.org/10.1007/s13369-025-09984-0
6. Gayathiri, B., & Ahamed, S. B. I. (2025). Assessing consumer perceptions of electric vehicles in India through the lens of the theory of planned behavior. *Journal of The Institution of Engineers (India): Series A*. https://doi.org/10.1007/s40030-024-00863-9
7. Campino, J., Mendes, F. P., & Rosa, Á. (2023). The race of ecological vehicles: Consumer behavior and generation impact in the Portuguese market. *SN Business & Economics*, 3(8). https://doi.org/10.1007/s43546-023-00524-2
8. Javadnejad, F., Jahanbakh, M., Pinto, C. A., & Saeidi, A. (2023). Analyzing incentives and barriers to electric vehicle adoption in the United States. *Environment Systems and Decisions*, 44(3), 575–606. https://doi.org/10.1007/s10669-023-09958-3
9. Shakya, L. K., Devkota, N., Dhakal, K., Poudyal, R., Mahato, S., Paudel, U. R., & Parajuli, S. (2024). Consumer's behavioural intention towards adoption of e-bike in Kathmandu valley: Structural equation modelling analysis. *Environment, Development and Sustainability*. https://doi.org/10.1007/s10668-024-04595-5
10. Madeira, C. (2024). Indebtedness and labor risk sorting across consumer lender types in Chile. *Journal of Banking Regulation*. https://doi.org/10.1057/s41261-024-00250-1
11. Pal, S. (2024). Do economic globalization and the level of education impede poverty levels? A non-linear ARDL approach. *The Journal of Economic Inequality*, 22(3), 621–667. https://doi.org/10.1007/s10888-023-09608-3
12. Mondal, S., & Suman, N. K. (2025). Challenges in embracing electric vehicles: Exploring the policy landscape and way forward. *Transportation in Developing Economies*, 11(2). https://doi.org/10.1007/s40890-025-00248-y
13. Zhang, L., Li, S., Cheng, X., & Geng, J. (2024). Can subsidies promote electric vehicles' sustainable development? A general equilibrium perspective on substituting enterprises for government. *Annals of Operations Research*. https://doi.org/10.1007/s10479-024-06402-w
14. Trung, T. T., Ngan, D. H., Nam, N. H., & Quynh, L. T. T. (2024). Framework for measuring high school students' design thinking competency in STE(A)M education. *International Journal of Technology and Design Education*, 35(2), 557–583. https://doi.org/10.1007/s10798-024-09922-5
15. Janamala, V., Rani, P. S., Rani, K. R., & Swarnasri, K. (2025). Transformative impact of electrical engineering on society, education, academia, and industry: A brief review. *Electrical Engineering*. https://doi.org/10.1007/s00202-025-03066-y
16. Pandita, D., Altekar, S., & Vapiwala, F. (2025). Are we willing to embrace the future? The need for moving towards sustainable mobility. *Discover Sustainability*, 6(1). https://doi.org/10.1007/s43621-025-01303-9
17. Bocken, N., Coffay, M., & Dalhammar, C. (2025). The Brussels and California effects? Circular economy policy influence across borders. *Circular Economy and Sustainability*, 5(3), 1665–1688. https://doi.org/10.1007/s43615-025-00516-4

Chapter 15

Global Perspectives

15.1 CULTURAL PERCEPTIONS OF ELECTRIC VEHICLES

This chapter concludes with a summary of electric vehicle (EV) technology, applications, infrastructure, market dynamics, and future possibilities [1]. EVs are changing global transportation ecosystems, and this chapter goes beyond a summary. This chapter mixes technical, environmental, and economic problems to analyze how significant actors will shape sustainable and smart mobility in the near future.

Technological Convergence and Innovation

The technological aspects of this chapter conclude that EVs represent a convergence [2–5] of numerous advanced systems evolving rapidly. EV hearts go beyond motors and batteries. Moreover, it links power electronics, energy storage, materials science, digital control, and networked infrastructure. A deregulated and digitally monitored vehicle architecture with upgraded lithium-ion batteries, regenerative braking, vehicle control units, and fast-charging cooperate to power the car, control energy flow, predict defects, maximize range, and provide real-time diagnostics.

EV propulsion is modular and scalable. Internal combustion engine (ICE) cars' mechanical complexity limits them. Batteries, bidirectional charging for vehicle-to-grid (V2G) applications, and software-driven performance upgrades can be added to EVs. Table 15.1 compares ICE and EV technology updates.

The table indicates that the beneficial side of EVs is to relieve themselves from the mechanical architecture while allowing digital optimization, paving the way for constant post-sale upgrades and integrations with intelligent mobility systems. The biggest technology revolution is shift from analog to software-defined cars.

DOI: 10.1201/9781003608097-15

Table 15.1 Comparison of Key Technological Attributes: ICE Versus EV Platforms.

Feature Category	*Internal Combustion Engine Vehicles*	*EVs*
Propulsion system	Engine with fuel combustion	Electric motor with battery power
Energy storage	Fuel tank (gasoline/diesel)	Battery pack (lithium ion, solid state)
Efficiency (tank-to-wheel)	~20–30%	~70–90%
Maintenance complexity	High (oil, belts, exhaust systems)	Lower (fewer moving parts, less wear)
Software integration	Limited	High (BMS, VCU, OTA updates)
Emission control systems	Required	Not applicable (no tailpipe emissions)
Platform flexibility	Low	High (shared architectures, modular systems)

Environmental Impact and Sustainability Metrics

EVs reduce local pollution, improve energy efficiency, and promote lifecycle sustainability. EVs have a smaller lifecycle carbon footprint than internal combustion engine vehicles (ICEVs), but battery production requires a lot of raw materials and energy, and the inclusion of production to phase management.

In metropolitan areas with high car density, EVs reduce greenhouse gas (GHG) emissions. When paired with renewable energy, electricity makes EVs fossil fuel-free. Table 15.2 presents a comparison of the lifecycle emissions of ICEVs and battery electric vehicles (BEVs).

Table 15.2 shows that BEVs have greater raw material extraction and manufacturing initial emissions but greatly reduced use-phase emissions. EVs using renewable charging reduce emissions by approximately half compared to ICE cars. The findings support EVs' importance in fulfilling national climate goals and improving urban air quality.

Market Adoption and Policy Support

Policy, economic incentives, and societal norms influence EV adoption. The market dynamics and policy framework chapters show that government support encourages early EV adoption. EV adoption has increased due to purchasing incentives, tax rebates, zero-emission norms, and public charging infrastructure.

However, regional adoption rates vary greatly. Consumer behavior and fleet composition depend on income, fuel prices, local energy grid

Table 15.2 Estimated Lifecycle GHG Emissions of ICEVs Versus BEVs (in Metric Tons CO_2e).

Lifecycle Stage	*ICEV (Compact)*	*BEV (Compact)*
Raw material extraction	1.2	3.5
Manufacturing	4.5	6.2
Use phase (per 150,000 km)	28.0	5.5 (renewable energy-powered)
End-of-life	0.8	0.9
Total emissions	**34.5**	**16.1**

Table 15.3 Market Share of EVs in Selected Countries (2024 Estimates).

Country	*EV Market Share (%)*	*Key Government Initiatives*
Norway	82	Tax exemptions, free parking, toll waivers
China	35	Subsidies, EV quotas, charging infrastructure
Netherlands	30	Company car tax benefits, city bans on ICE
United States	12	Federal tax credit, state-level incentives
India	2.5	FAME-II scheme, state subsidies, charging grants

composition, and urbanization. Table 15.3 shows inconsistencies and policy links in EV market shares of select countries.

Table 15.3 shows that countries with coordinated policy efforts have achieved higher adoption rates [6,7]. Norway's success ties to multilayered incentive schemes and restrictions on ICEVs, while in China, industrial policies help push local manufacturing. Emerging economies are yet in the early stages, constrained by infrastructure and affordability.

Infrastructure and Charging for Operational Feasibility

On various accounts, a recurring problem discussed in many chapters is the construction of a fool-proof charging infrastructure. While home charging is convenient for individuals, a dense and interoperable public charging network is essential for mass adoption. On the other hand, fast-charging technologies are necessary to match the refueling experience of conventional vehicles and support the high-utilization segments of taxis, logistics fleets, and long-distance travel.

The charge, related to infrastructure development, is dependent on grid capacity planning and broader urban planning strategies. With the growing numbers of EVs, local distribution networks are stressed, providing the opportunity for smart charging, load balancing, and integration with renewables. Table 15.4 summarizes the types of charging solutions, average power delivery, and target applications.

Table 15.4 demonstrates the stratification of charging needs. Home and workplace charging generally addresses everyday local commutes. The charging infrastructure characterized by DC fast and ultra-fast stations targets only high-throughput or long-distance applications. Good infrastructure design helps EV ecosystems scale.

Future Integration and Industry Transformation

This book's ultimate and most important discovery is that electric cars can alter mobility. The development of intelligent, decentralized transportation systems will be enabled by the integration of EVs with autonomously behaving systems, frameworks of connected vehicles, and energy distribution networks. Thus, a desirable future perspective includes many advanced features and even earlier deployable features such as V2G energy exchange, predictive maintenance through telematics, artificial intelligence (AI)-driven fleet optimization, and monetizing data from onboard sensors. Eventually, EVs will be part of the future of mobility-as-a-service (MaaS) models, which move transportation toward a subscription-based or shared service model rather than focusing on personal vehicle ownership, as defined in Table 15.5. Table 15.5 provides the projected strategic roles of EVs in the future mobility ecosystem by 2035.

15.2 SOCIAL IMPACTS OF ELECTRIC VEHICLE ADOPTION

EVs are not stand-alone products but dynamic nodes in a digital, environmental, and logistical ecosystem. The roles and functions of EVs are poised to grow over time with improvements in software, power systems, and urban

Table 15.4 Overview of EV Charging Types and Use Cases.

Charging Type	*Power Range*	*Charging Time (Compact EV)*	*Use Case*
Level 1 (AC)	1–2 kW	8–20 hours	Home charging, low mileage users
Level 2 (AC)	3–22 kW	3–8 hours	Residential, workplace, public
DC fast charging	50–150 kW	30–90 minutes	Highway corridors, fleet depots
Ultra-fast charging	150–350+ kW	15–30 minutes	Intercity travel, commercial use

Table 15.5 Strategic Roles of EVs in the Future Mobility Ecosystem (Projected 2035).

Role	*Description*	*Potential Impact*
Zero-emission transport	Replacement for ICEVs	Air quality, climate change mitigation
Energy storage node	Bidirectional charging with grid support (V2G)	Grid stability, renewable integration
Data generation platform	Real-time sensors and connectivity	Traffic optimization, insurance analytics
Autonomous shuttle base	Integration with self-driving systems	Urban redesign, mobility access
Circular economy enabler	Battery recycling and remanufacturing infrastructure	Resource conservation, cost efficiency

policies, rendering them ever more critical. As mentioned in the previous chapters, technology transition, environmental transition, economic transition, and even social transition all cite EVs as the crowning jewel of the transition process. Electric propulsion has brought energy efficiency, lower emissions, and better, greener, and fairer transportation systems. Naturally, affordability, infrastructural, and raw material issues exist.

Lifecycle Emissions Perspective

Despite many obstacles, electric mobility is growing. Governments, industry stakeholders, researchers, and consumers must collaborate to optimize EV benefits across use cases and locales. Electric cars will provide the framework for replacing combustion engines and transportation infrastructure in the coming decades.

GHG emissions arise from fossil fuel combustion in ICEs, mainly due to transportation. EVs are being promoted as a way to decarbonize transport and combat climate change. This section provides an evidence-based vision of zero-GHG EV manufacturing, operation, and end-of-life management. Also, national energy strategies, air quality in urban soil, and global emission reduction commitments are discussed in this section, and all these would capture the overall systemic impact of wide-spread EV deployments. One of the most important contributions of EVs to GHG reduction emerges when viewing their total emissions over the product lifecycle. One of the often cited claims against EVs is that their development requires an inordinate amount of energy and that EVs produce more CO_2 than ICEVs. However, a proper lifecycle emission assessment on raw material extraction, vehicle manufacture, energy consumption in operation, and vehicle disposal reveals that an EV has a net reduction in GHG emissions throughout its usable life.

Use-phase reductions predominate, as carbon dioxide emitted through the combustion of fuel is released directly as tailpipe emissions from ICEVs. For these reasons, battery EVs have no tailpipe emissions. Instead, their operational emissions depend on the type of electricity used for charging. If the energy supplied is predominantly from renewable resources—such as hydraulic, wind, or solar energy—the operational carbon footprint of an EV can approach zero. But even in regions where electricity partially comes from fossil fuels, high efficiencies of EV powertrains—usually three to four times that of ICEVs—will deliver lower emissions per kilometer traveled.

Over time and with increasing mileage, the cumulative GHG advantage of EVs becomes more visible. Extensive modeling exercises and studies have demonstrated that an upper mid-sized EV achieves comparable emissions to a similar ICEV within 30,000–50,000 km. From this point onward, savings start to accumulate. This is particularly pertinent for fleet vehicles, public transport, or delivery vans, which accrue high mileage over a short period, which could positively impact the environmental payback of an electric driveline investment.

The Source of Energy and Carbonization of the Grids

EVs' emission benefits derive from their electrical supply's carbon intensity and their use. Thus, EV adoption's climate benefits depend on energy grid decarbonization. In fully decarbonized energy networks with primarily renewable electricity, EVs emit almost nothing indirectly. Due to electric drivetrain efficiency, EVs reduce carbon emissions more than ICEVs, even on a partially decarbonized grid.

Energy Source and Grid Decarbonization

Meanwhile, from end-use emissions, where the charging sources are linked to CO_2 emissions and thus to the global warming potential of the electricity used, feeding these vehicles with cleaner energy sources is very important in maximizing their climate value. While in the fully decarbonized energy systems the energy for generating electricity is drawn mostly from renewable power sources, EVs generally function with almost no indirect emissions. Partially decarbonized networks will have EVs functioning with reasonable reductions compared to the emissions from ICE automobiles because of the efficiency improvement in electric drivetrains.

Vehicles like these make deeper decarbonization of the grid possible with features such as V2G schemes. EVs can then act as distributed energy storage units consuming bidirectional power—between the grid and the EVs. Flexible storage systems balance supply and demand for intermittent renewables like solar and wind energy. Large EV fleets can reduce renewable energy curtailment and provide grid services like frequency management and peak shaving if managed well. While synergies reduce GHGs, EVs and clean electricity feed each other.

While reducing CO_2 emissions, EVs also cut NOx and PM2.5 emissions, which result from poor urban air quality and high traffic density. Although not GHGs, these pollutants indirectly affect climate forcing and serious public health consequences. Therefore, electric car emission reductions benefit the world and local communities.

Manufacturing Footprint and Material Contemplations

Most GHG savings occur with EV operation; however, lithium-ion battery manufacturing requires consideration. This energy-intensive battery uses lithium, cobalt, nickel, and graphite. The carbon footprint of battery production depends heavily on the energy source used in the production facility. For example, manufacturing batteries in regions with coal-based electricity grids leads to significantly higher emissions compared to that in facilities powered by hydroelectric or solar energy.

Lifecycle assessments have shown that despite increased upfront emissions, the amount of emissions over an EV lifespan remains much lower compared to that of ICEVs. Gradual improvement in battery energy density, manufacturing efficiency, and recycling practices results in reduced emissions related to EV production. For example, moving from cobalt-rich chemistries to alternatives such as lithium iron phosphate (LFP) reduces dependence on energy- and resource-intensive materials. Similarly, investments in battery recycling infrastructure enable the recovery and reuse of metals, thus reducing the need for virgin material extraction and associated emissions.

The integration of closed-loop recycling systems for EV batteries makes electric mobility even more favorable in terms of emissions assessment. Governing the models of circular economy are such practices as recovering materials from the end-of-life batteries and reintroducing them into new cell production, thus achieving considerable carbon intensity reductions in battery manufacture. Now there are several large manufacturing companies, as well as various research institutions, working toward the cost-effective and environmentally friendly enhancement of these recycling processes.

Segment Transformation and Fuel Replacement

EVs have considerable potential to disrupt the transportation sector's dependency on fossil fuels, especially those sections that use gasoline and diesel as fuels, which are among the largest sources of anthropogenic carbon dioxide emissions. Thus, even with a moderate uptake of EV technology in major markets by the end of 2040, the International Energy Agency (IEA) has estimated millions of barrels a day to be displaced from the current numbers of crude oil consumption. Less fossil fuel consumption, in turn, directly relates to less GHG emissions in those areas, for example, affecting oil extraction, refining, and distribution.

This displacement effect will be most effective when combining policies that speed up the introduction of EVs and mandate a reduction in fossil fuel vehicle sales. Countries' phased bans on new ICEVs in terms of sales, renewable energy objectives, and emission reductions set at Paris Agreement levels all come into play concerning this scenario. EVs have become very diverse and have played a significant role in achieving these nationally determined contributions (NDCs) regarding a key factor: electrification is the part closest to decarbonizing transport, a sector that is difficult to decarbonize relative to others.

The use of EVs also extends beyond passenger cars because electrifying public transport, commercial fleets, and logistics operations would be critical GHG mitigation strategies. Because electric buses, trucks, and delivery vans mostly run on fixed routes and high-mileage cycles, they maximize battery use and deliver significant emissions savings. Fleet electrification programs have shown significant reductions in transport-related emissions in cities like Shenzhen, Los Angeles, and London and are already being scaled up for application in other urban centers.

Temporal and Geographic Variability in Emission Benefits

The GHG benefits from using an EV are not uniform in every context. They depend on a number of differing factors like energy mix of the localized region, type of vehicle, way it is used, and availability of infrastructural facilities in different countries. With regard to coal-fired electricity generation in Poland or in parts of India, it is possible that an EV produces higher emissions when operational, compared with countries having cleaner grids, like Norway or Canada. However, even within coal grids, most emit less due to better efficiency from electric drive trains.

In addition, temporal factors are intertwined with the emissions signature of EVs. With the national grids decarbonized, the indirect emissions from charging EVs are reduced over time, even for those EVs already on the road. ICEV emissions remain constant or increase with age and wear, but grid clean-up increases.

Using patterns matters. High mileage rate reduces lifetime GHG and amortizes quickly. In contrast, low-mileage-use cases may gradually reduce emissions. Even without tailpipe emissions, carbon accounting often fails to recognize air quality gains.

EVs will reduce GHG emissions with supportive legislation and strategic planning. Electric car incentives, renewable energy infrastructure investment, and battery production and recycling requirements must be part of national and regional transport, energy, and industrial agendas.

Cap-and-trade or carbon taxes can internalize ICEV emission costs, helping EVs compete. Fuel economy and zero-emission vehicle laws promote low-carbon mobility.

Strategic planning should distribute EV technology benefits fairly. Many low-income and rural areas have access constraints because to higher upfront costs and lack of charging infrastructure. EV GHG benefits will spread by supporting vehicle affordability, electrifying public transport, and creating easy-entry charging networks statewide.

Changes to the Automotive Industry

To produce electric vehicles, the global auto industry is changing drastically. Electrification, digitization, and sustainability are upending a century-old paradigm built on ICEs, mechanical complexity, and fossil fuels. This revolution changes automotive business models, supply networks, and engineering processes—competition dimensions—in addition to replacement technologies. This shift toward electrification, connection, automation, and shared mobility will change personal and commercial transportation.

From Internal Combustion to Electrification

The biggest change is switching from ICEs to battery electric powertrains. Switching from engine-driven to electric powertrain architectures requires rethinking automotive engineering. Simpler mechanical systems rely more on software and high-efficiency energy storage. Modular platforms based on electric motors, inverters, and battery packs have replaced fuel-driven powertrains, gearboxes, and emission control systems. Less mechanical maintenance and fewer moving parts improve vehicle durability.

The transition to electric mobility is disruptive enough for automakers. Legacy manufacturers are trying to redefine technical knowledge and industrial capacities after decades of ICE development. Retooling factories and studying battery, thermal, and power electronics are needed. Technology-driven automotive startups use digitally native business models and vertically integrated production. Most companies base their retail and aftersales on direct-to-consumer, redesigned models, not dealerships.

However, the shift will cost enormous capital. Factories are investing billions in EV development, gigafactories, and strategic partnerships with battery and software businesses to supply users with electricity. Certain expenditures ended ICEV manufacture within a decade. Such commitments signal an electrified mobility future because delaying the shift could make companies outdated in fast-changing competitions.

Disruption of Supply Chains and Value Chains

Electric automobiles impact automotive supply chains. Engines, exhaust systems, and gearboxes used traditional supply lines, but lithium-ion cells, battery management systems, electric motors, and semiconductors are popular. This initiative requires thousands of vendors to transition to EV supplies or risk closure.

The transition to EVs adds geopolitics to automobile supply chains. Vehicle manufacture requires lithium, cobalt, nickel, and graphite mining. Regional concentration makes these materials vulnerable to global resource security rivalry. Automotive companies invest in upstream mining, negotiate long-term supply agreements, and promote recycling to reduce virgin extraction. The automobile industry aims to include battery supply chain traceability and ethical sourcing, supporting environmental and human rights collaboration.

This transitions the automotive supply chain from linear to circular. Long-term product initiatives include battery recycling, energy storage, and closed-loop material recovery. This would help firms satisfy environmental and regulatory targets while lowering costs and supply concerns. Circular practices change vehicle lifecycle value capture, manufacture, and retention.

Integration of Software and Digital Services

Software and digital services are also emphasized in vehicles. EVs simplify car mechanics, enabling digital breakthroughs. Centralized computing, OTA software updates, and telematic connectivity provide real-time monitoring, predictive maintenance, and data-driven performance optimization in current EVs.

Software's growing prominence changes industry competition. Instead of hardware, automakers may offer software-defined platforms. Smart charging analytics, remote diagnostics, advanced driver-assistance systems (ADAS), and infotainment subscriptions can bring in revenue. The auto industry is moving beyond production and dealerships to platform-based economics and digital service ecosystems.

Changes impact talent needs. Automotive companies now compete for software engineers, data scientists, and user experience designers, which were never their strengths. Change is shown by technological partnerships, mobility startup purchases, and software development divisions. Connected and automated EVs generate, retain, and transmit sensitive user and operational data, raising cybersecurity and data privacy concerns.

Cars with software allow additional customization. Users can alter car configurations, infotainment interfaces, and driving modes via digital interfaces, while manufacturers can expand vehicle capabilities using OTA updates. Like smartphones, this continuous improvement strategy changes customer expectations for automobile ownership and operation.

New Business Models and Market Dynamics

Car electrification is driven by technology and new business models without ownership rules. EVs integrate with smart infrastructure, urban mobility systems, and shared transportation networks, changing automotive sales to services.

Mobility-as-a-service (MaaS) platforms, ride-hailing networks, and subscription vehicle access models are developing in metropolitan areas where automobile ownership is expensive and onerous. Electric automobiles are popular for their lower running costs, easier maintenance, and digital connectivity. Fleet operators prefer EVs for shared mobility due to asset utilization and long-term cost efficiency. EVs transform current transportation and delivery.

Low power costs relative to gasoline or fuel, maintenance capital savings, and EV TCO improvement throughout operations are convincing. This process affects environmentally conscientious customers and cost-conscious businesses. Government incentives, fuel savings, and minimum emission fines are boosting EV sales in cost-sensitive areas.

The key themes include car industry political motivations. Emissions, zero-emission vehicle, and fuel economy criteria are pressuring automakers into rushed EV research and risk stiff financial penalties. Policymakers in various nations have set ICE automobile sales phaseout dates. This announcement gives industry planners a deadline and supports clean transportation technology investment.

Heavy investment is achievable with long-term electrification road networks, specific EV platforms, and automotive joint ventures to spread development costs and scale. Legal obligations aligning with corporate goals have fostered extraordinary innovation and funding. Public–private partnerships, national battery alliances, and EV-focused industrial policies shape the business.

Environmental, social, and governance (ESG) compliance is becoming mandatory for global automakers. Investors, consumers, and regulators want carmakers to disclose carbon emissions for anti-bribery checks. The disruption of technology by electrification can affect company sustainability strategies, brand equity, market access, and shareholder values.

REFERENCES

1. Zhao, Y., & Xu, W. (2023). Measurement of risk spillover effect based on EV-Copula method. *Humanities and Social Sciences Communications*, 10(1). https://doi.org/10.1057/s41599-023-02287-5
2. Yang, C., & Mu, X. (2023). Mapping the trends and prospects of battery cathode materials based on patent landscape. *Frontiers in Energy*, 17(6), 822–832. https://doi.org/10.1007/s11708-023-0900-x
3. Podgorbunsky, A. B., Opra, D. P., Zheleznov, V. V., Sinebryukhov, S. L., &Gnedenkov, S. V. (2025). Hard carbon for sodium-ion batteries: Progress and prospects for application. *Russian Journal of Inorganic Chemistry*, 70(2), 198–213. https://doi.org/10.1134/s0036023624603866
4. Rawat, S. S., Kumar, R., & Das, K. (2024). Lithium-ion battery progress in surface transportation: Status, challenges, and future directions. *Multimedia Tools and Applications*. https://doi.org/10.1007/s11042-024-20505-3

5. Paul, D., Nandi, A., Bhattacharyya, S., Das, S., & Biswas, N. (2025). A comprehensive review on lithium-ion battery thermal management (BTM) using phase change materials: Advances, challenges, and future perspectives. *Journal of Thermal Analysis and Calorimetry*. https://doi.org/10.1007/s10973-025-14260-2
6. Mystakidis, A., Tsalikidis, N., Koukaras, P., Skaltsis, G., Ioannidis, D., Tjortjis, C., & Tzovaras, D. (2025). EV charging forecasting exploiting traffic, weather and user information. *International Journal of Machine Learning and Cybernetics*. https://doi.org/10.1007/s13042-025-02643-8
7. Singh, V. (2025). IoT-enabled load power factor optimization in fast EV chargers. *Journal of Umm Al-Qura University for Engineering and Architecture*. https://doi.org/10.1007/s43995-025-00177-2

Chapter 16

Emerging Technologies

16.1 SOLID-STATE BATTERIES

Battery technology evolution is an important factor in determining the future of EVs; among potential innovations, solid-state batteries (SSBs) are perhaps the most revolutionary advancement. SSBs differ from the standard lithium-ion batteries in that they substitute the liquid electrolyte with solid electrolytes, thus granting them dozens of performance, safety, and design advantages. In this chapter, we will analyze the scientific principles, material novelizations, technical advantages, manufacturing and fabrication challenges, and further potentials of SSBs in the context of EVs.

Overview and Scientific Principle

SSBs fundamentally differ from conventional lithium-ion batteries in their internal structure. In a conventional lithium-ion battery, the electrolyte is a liquid that, during charging and discharging, facilitates the transfer of lithium ions between the anode and the cathode. Because of the safety issues associated with the use of liquid electrolytes, especially leakage and flammability, battery fires and explosions have become not too uncommon. In SSBs [1–3], the liquid can be eliminated and replaced by solid electrolytes such as ceramics, glasses, or polymer-based conductors, removing most of the safety concerns with contemporary battery technologies.

A SSB consists of three essential parts: a lithium-metal or lithium-alloy anode, a solid electrolyte separator, and a cathode with a high energy density. The interface between the solid electrolyte and the electrodes is of utmost importance, where ion conduction must take place across the boundaries. SSBs are expected to provide higher energy densities, lower charge time, and better lifecycle stability. These features are particularly attractive for the automotive sector, which desires compact, safe, and durable energy storage systems.

DOI: 10.1201/9781003608097-16

Comparative Performance Metrics

Showing the advantages of "solid state" against liquid-type lithium-ion cells, a tabulation of key technical parameters is provided in Table 16.1.

This comparison highlights some key advantages. First, SSBs have almost two times the energy density, which allows EVs to run for longer distances on a single charge. Second, SSBs enjoy significantly longer cycle life, so their replacement is further apart over the lifetime of an EV. Third, fast charging, which is one of the main challenges for widespread EV adoption, appears more plausible for SSBs due to lower internal resistance and better thermal stability. Last, the removal of combustible liquid electrolytes greatly enhanced the safety profile of an EV, an approach of which consumers are increasingly concerned, especially with battery fires now being an acute worry.

Materials and Electrolytes Chemistry

Material selection is the most important variable with regard to performance in SSBs. Solid electrolytes can be divided into three broad categories: oxide, sulfide, and polymer. Each type has its advantages and disadvantages, as summarized in Table 16.2.

Table 16.1 Comparative Performance of Lithium-Ion Batteries versus SSBs.

Parameter	*Conventional Li-Ion Battery*	*SSB*
Energy density (Wh/kg)	150–250	300–500
Cycle life (charge cycles)	1,000–2,000	2,500–5,000
Charging time (0–80%)	30–60 minutes	10–20 minutes
Operating temperature range	−20°C to 60°C	−30°C to 100°C
Flammability risk	High	Negligible
Cost per kWh (est.)	$120–150	$70–100 (projected)

Table 16.2 Types of Solid Electrolytes in SSBs.

Electrolyte Type	*Ionic Conductivity (S/cm)*	*Mechanical Strength*	*Chemical Stability*	*Process Complexity*
Oxide-based	10^{-4} to 10^{-3}	High	High	High
Sulfide-based	10^{-3} to 10^{-2}	Medium	Moderate	Moderate
Polymer-based	10^{-5} to 10^{-4}	Low	Low	Low

Oxide electrolytes like lithium lanthanum zirconium oxide (LLZO) are thick and chemically stable, but they are brittle and require high-temperature soldering, which is complicated to manufacture. Sulfide materials, such as lithium germanium phosphorus sulfide (LGPS), offer the highest ionic conduction and easy processing but are highly susceptible to moisture and need encapsulation. Polymer electrolytes are easy to handle and flexible but generally yield poor conductivity at room temperature, limiting their use in advanced electric vehicle (EV) batteries.

Material research in this area is heavily concentrated on enhancing ionic conductivity while maintaining chemical compatibility with both the anode and the cathode. Techniques to reduce interfacial resistance and maximize battery life durability are based on advanced coatings of materials, composite electrolyte structures, and engineering interfaces.

Manufacturing Challenges and Scalability

The promise of SSBs is well acknowledged; nevertheless, the transition from lab-scale prototypes to commercial-scale manufacture is nontrivial. The manufacturing processes of SSBs require strict control of material purity, layering thickness, and bonding on the interface. Table 16.3 compares the manufacturing complexities between lithium-ion batteries and SSBs.

SSB production often involves processes like tape casting, hot pressing, and pulsed laser deposition. With the liquid electrolyte flowing and self-leveling in conventional batteries, the solid electrolyte has to rely on some sort of mechanical or thermal processing to achieve good contact with the electrodes. The retention of interfacial integrity without delamination for repeated charge and discharge cycles is one of the key engineering challenges associated with the SSB manufacturing process.

In addition, interfacing with lithium metal anodes presents an additional challenge due to the aggressive handling, owing to its reactivity with moisture and oxygen. The presence of lithium metal anodes thus translates to controlled dry room environments [4–6], with highly controlled humidity, which adds to capital expenditure for battery manufacturers.

Table 16.3 Manufacturing Considerations for SSBs Versus Li-Ion Batteries.

Parameter	*Lithium-Ion Battery*	*SSB*
Electrolyte application	Liquid filling (simple)	Thin-film or sintering
Anode material	Graphite (stable)	Lithium metal (reactive)
Interface control	Minimal	Critical for ion transport
Dry room requirements	Moderate	High (especially for sulfides)
Capital expenditure	Established and optimized	High due to novel equipment

Safety Improvements and Thermal Stability

SSBs offer a significant improvement with regard to safety over classical lithium-ion batteries. With the absence of flammable solvents being complemented by thermally stable electrolytes, the risks of thermal runaway—a sort of chain reaction that may cause fires or explosions in traditional cells—are greatly decreased. SSBs ameliorate the regular safety concerns associated with lithium-ion technology, as illustrated in Table 16.4.

From one year to another, one persistent safety concern in lithium-ion batteries has been that of dendrite formation. Dendrites are long, needle-like structures made of lithium that can sometimes grow during charging and pierce a separator to create an internal short circuit. Solid electrolytes, especially ceramic and sulfide ionic types, act as physical barriers, suppressing growth propagation in dendrites to enhance the safety aspect of the batteries during operations.

Thermal runaway is the phenomenon that usually happens due to overcharge and external heating, and it is among the fastest to escalate in a liquid electrolyte environment. Alternatively, SSBs improve the thermal stability by a wide margin. Their strength makes them favorable in a high-temperature environment and favorable for use in high-density compact battery packs where heat dissipation is relatively low.

Applications in Automotive Platforms

The impact on the automotive sector from adopting SSBs can be understood in a nutshell. High energy density provides a greater range per charge, while a faster charging solution makes refueling shorter, both of which directly affect consumer acceptance. Many auto manufacturers are currently investing heavily in SSB research and pilot production lines. Table 16.5 presents a high-level summary of many of the world's leading manufacturers and their activity in SSB ventures.

These timelines embody not only the expectation but also the cautiousness regarding technology within the industry. For most of the companies, the belief is that SSBs will first penetrate the high-end or performance EV markets, with wider acceptance to follow once the costs associated with

Table 16.4 Safety Comparison of SSBs and Lithium-Ion Batteries.

Safety Parameter	*Lithium-Ion Battery*	*SSB*
Electrolyte flammability	High	None or very low
Thermal runaway risk	Moderate to high	Low
Internal short circuit risk	Present (dendrite growth)	Reduced with solid barrier
Gas evolution during failure	Yes	Minimal or none

Table 16.5 Major Automotive Companies and SSB Initiatives.

Automaker	*Partner/Technology Provider*	*Target Year for Deployment*	*Battery Type Focused*
Toyota	Panasonic, QuantumScape	2027–2030	Sulfide-based SSBs
BMW	Solid Power	2026–2028	All-solid-state batteries
Volkswagen	QuantumScape	2025–2026	Ceramic-electrolyte SSBs
Nissan	In-house R&D	2028	Polymer–ceramic hybrids
Hyundai	SolidEnergy Systems	2027	High-energy lithium metal

production and volume become less prohibitive. In summary, automotive applications require not only high energy density but also high power output, long cycle lives, and, above all, safety—all criteria that SSBs are increasingly being able to satisfy as the technology matures.

Long-Term Forecast and Scopes of Research

The future of SSBs depends on the overcoming of a few remaining scientific barriers and engineering obstacles. The promising active fields of current research are increasing ionic conduction at room temperature and developing composite electrolytes with a combination of flexibility and strength and optimizing interfaces at lithium–metal anode interfaces. Research institutions and private companies experiment with nanostructured materials and hybrid electrolytes in an attempt to push the performance margin further.

Long-term SSBs could also be used in novel EV architectures; the compact form factor of the SSB permits battery pack designs that can be integrated into the structures of vehicles (structural batteries). The higher energy density and thermal safety also permit new applications, such as electric aviation and grid-level storage.

The developments in SSBs will most likely be followed by promising environmental outcomes. Many solid electrolytes are less environmentally hazardous than their liquid counterparts [7–9]; coupled with the usually higher cycle life associated with these batteries, reduced disposal frequencies can thus be expected. SSB material recycling strategies are still in their early development stages, while supply chain sustainability remains a focus of interest.

In summation, SSBs are a significant and progressive breakthrough in energy storage technology under the umbrella of EVs. SSBs address many limitations posed by conventional lithium-ion systems, such as limited range, long charging times, and safety threats, thereby making it possible to accelerate global transitions in electrified transportation. However, many technical and economic hurdles remain; nevertheless, rapid innovation, heavy investments, and large demand in the automotive sector make SSBs one of the most promising and relevant areas within the future for electric mobility sets.

16.2 WIRELESS CHARGING

Currently, wireless charging technology represents one of the most revolutionary advances in the field of electric mobility, offering new dimensions of convenience, automation, and lifecycle integration to EV infrastructure. With the surge in EVs, the growing need for simple, user-friendly, and efficient infrastructure for charging has become crucial. There is a new approach introduced by wireless or inductive charging in the way that EVs could be powered: no more cables, less physical wear and damage, and dynamic and automated energy transfer systems that could easily be installed in the built environment. This section examines the fundamentals of wireless charging, the methods and materials, the key performance parameters, the obstacles to deployment, and efficiencies for comparison and future scopes.

Fundamentals of Wireless Charging

Wireless charging is basically electromagnetic induction that is used for charging EVs. In a typical system, charging pads are embedded in the ground to transfer energy to the receiver coils mounted to the vehicle itself. A primary coil from the energized power grid in the ground generates a fluctuating electromagnetic field. The current transfers to the secondary coil on the EV, charging the battery for consumption. This basically operates like the mouth of a transformer, but here the air acts as the medium instead of a ferromagnetic core in different scenarios.

Inductive charging is the preferred method, but other methods such as resonant inductive coupling and capacitive coupling have been studied. Resonant systems can tune the transmitter and receiver at the same resonant frequency to transfer energy with high efficiency over longer distances. Capacitive coupling uses electric instead of magnetic fields, which is still in a largely experimental phase for EVs due to problems of misalignment and low energy transfer efficiency.

Wireless EV charging systems are generally designed with power ranges from 3.7 kW up to more than 100 kW for high-performance applications. Such devices are intended to work in a few inches of air gap between a transmitter and a receiver [10–12], with the ability to tolerate minor misalignments since efficiency can be enhanced by properly aligning the two systems. Obviously, the most appealing aspect of wireless charging is that it can pave the way to modernizing charging models, such as dynamic charging, where a vehicle gets energy even in a moving state.

Technical Components and System Architecture

Wireless charging systems for EVs usually have three major subsystems: the ground assembly (transmitter), the onboard assembly (receiver), and the control and communication unit. The configuration of the transmitter coil integrated into parking surfaces and roadways is meant to generate an alternating magnetic field. The receiver coil, which is highly tuned to capture the transmitted energy, is mounted below the vehicle chassis. A power control unit manages the power flow, safety, and communication protocols. Table 16.6 presents the various components of a wireless EV charging system and their functions.

All the components mentioned in Table 16.6 should work harmoniously while following standardized protocols in order to guarantee the performance and interoperability of various vehicle types and brands. An inbuilt communication system in the EV carries out preliminary checks before initiating the actual charging session in the vehicle. Such alignment technologies, which mainly comprise a magnetic field mapping and visual guidance system, are usually integrated for the benefit of consumers.

Efficiency and Performance Parameters

Historically, wireless charging systems have been inefficient compared to plug-in systems without an energy loss associated with electromagnetic transmissions. However, in the short time of their advancement [13–15], there has been a closing gap. Very well-aligned systems can now reach efficiency rates between 92 and 94%. Major performance parameters primarily consist of energy transfer efficiency, positional tolerance, power throughput, electromagnetic interference (EMI), and thermal management. Table 16.7 presents a comparison between wireless and wired charging systems regarding different parameters that are relevant to the operation of EVs.

Table 16.6 Key Components of a Wireless EV Charging System.

Component	*Location*	*Function*
Transmitter coil	Ground side	Generates alternating magnetic field using grid power
Inverter	Ground side	Converts direct current (DC) from grid to high-frequency alternating current (AC)
Receiver coil	Vehicle side	Captures magnetic field and induces current
Rectifier	Vehicle side	Converts induced AC to DC for battery storage
Alignment system	Both sides	Assists in positioning the vehicle relative to the coil
Communication module	Both sides	Facilitates data exchange for authentication/safety

Table 16.7 Comparison of the Performance of Wireless and Wired Charging Systems.

Parameter	*Wired Charging (AC/DC)*	*Wireless Charging (Inductive)*
Max efficiency (%)	95–98	90–94
Position sensitivity	Low	Moderate to high
Charging automation	Manual	Fully automated
Installation complexity	Moderate	High (especially embedded)
User convenience	Medium	Very high
Cost of infrastructure (per unit)	\$500–\$1,500	\$2,000–\$5,000

Wired charging systems are slightly more efficient even under the best conditions, but the convenience and speed of automation offered by wireless systems are indeed great. The elimination of manual handling of heavy cables, exposure to the weather, and corrosion make this technology particularly suitable for use in shared autonomous vehicles and high-frequency fleet applications. Such systems provide sealed and tamper-proof architecture, which again means much in terms of operational robustness in public spaces.

Deployment Scenarios and Use Cases

The most typical and commercially realistic scenario is static wireless charging, whereby a vehicle charges while parked over a ground pad. This is perfect for private home garages, public parking lots, bus depots, and taxi stands. The safety net for transit agencies is to use static wireless charging at endpoints or layover stops to conduct what could be called a "set and forget" approach in topping off bus batteries. The most interesting emerging applications include semi-dynamic charging, where vehicles charge at stops or even traffic lights.

Dynamic wireless charging is the most ambitious form [16–18], where energy is transferred from coils embedded in roadways to vehicles that are in motion. With such technology, driving range could become virtually unlimited, while the batteries onboard would become smaller, thus significantly reducing vehicle weight and cost. Pilot projects for dynamic charging have already started in a few cities and research centers. However, it requires a huge commitment of resources and complicated adjustment of urban infrastructure. Table 16.8 presents the different types of wireless charging systems according to deployment characteristics and technology readiness.

Static systems have already been commercially deployed, and data indicates that some companies, including WiTricity, Qualcomm Halo, and Momentum Dynamics, have proven operational systems. These can be

Table 16.8 Categories of Wireless Charging Systems.

Charging Type	*Power Range (kW)*	*Use Case*	*Technology Readiness Level*
Static charging	3.7–22	Home, office, public parking	High (commercially available)
Semi-dynamic charging	11–50	Bus terminals, taxi zones	Medium (pilot projects)
Dynamic charging	20–200	Highways, urban roads	Low (R&D and testbeds)

considered as compatible with the models of various vehicles and can be piloted in municipal fleets and for public sales. Dynamic systems are promising but face challenges regarding the current cost measure, standardization, and electromagnetic compliances.

Safety, Standardization, and Regulatory Framework

Safety is paramount because any system transfers energy between a high-active medium and a relatively low-electricity medium in public space. Wireless EV charging systems must comply with field exposure guidelines set by the International Commission on Non-Ionizing Radiation Protection (ICNIRP) and IEEE. Shielding materials and field-containment designs are used to prevent leakage and thereby ensure safety for people and others within the process.

Moreover, misalignment errors and detection of foreign objects (FOD) are two important operational issues. Wireless systems have sensors to find the presence of metallic or biological objects between the two ends of the charging pad and vehicle; this can cause energy loss if not taken care of in time, creating a safety risk. In addition, active closed-loop cooling is integrated with real-time power modulation systems in insulation against heat build-up during high-power transmissions.

The very existence of new state-of-the-art wireless charging technologies has resulted in the emergence of newer international standards, while maturing the existing state standards. The new wireless power transfer standard J2954 was published by the Society of Automotive Engineers (SAE) for transferring up to 11 kW of wireless power. The standard includes interoperability specifications, coil designs, and safety protocols. Table 16.9 shows the key regulatory standards associated with wireless charging technology.

The added value offered by wireless charging is disputed under economic considerations. At present, wireless systems cost more than conventional chargers due to coil alignment system complexities, embedded infrastructure installations, and shielding. But the application of large-scale production, standardization, and new materials is likely to bring down costs markedly in another ten years.

Table 16.9 Key Standards Governing Wireless Charging for EVs.

Standard	*Issuing Body*	*Description*
SAE J2954	SAE	Defines wireless power transfer protocols and interfaces
IEC 61980	IEC	Covers general requirements and safety of wireless power transfer (WPT) systems
IEEE 2030.1.1	IEEE	Focuses on AC/DC interfaces and interoperability
ISO 19363	ISO	Specifies safety aspects and magnetic field emissions

Table 16.10 Economic Impact and Market Forecast for Wireless Charging.

Metric	*2024 Estimate*	*2030 Forecast*	*CAGR (%)*
Global market size (USD)	$1.2 billion	$14.4 billion	~43%
Installed units worldwide	25,000	1.2 million	High growth
Average cost per system (USD)	$3,500	<$1,000 (projected)	Declining

There is the expected initial application of wireless charging in niche markets, with gradual acceptance into the mainstream consumer market: fleets, shared mobility, and premium EV segments. The economic outlook for wireless charging within the coming decade is shown in Table 16.10. Initial deployment is likely to occur in high-utilization EV segments before mainstream consumer uptake. Table 16.10 summarizes the projected economic impact of wireless charging over the next decade.

These figures underline the rapid expected expansion of the wireless charging market, particularly in urban centers—logistics hubs and autonomous vehicle ecosystems. The move toward contactless and automated solutions provides additional reinforcement to wireless technologies, especially as cities invest in intelligent transport systems.

Future Scopes and Integration with Emerging Trends

Wireless charging is likely to evolve into an enabling layer in smart transportation infrastructures. The application of automating cars seems especially promising as it obviates anything human in the charging process, lending itself perfectly to self-driven fleets and robo-taxis. Furthermore, the wireless systems can accommodate vehicle-to-grid (V2G) platform applications, which empower the vehicles in park to discharge energy back to the grid not through physical connectors but allowing bidirectional energy flow seamlessly.

Another promising direction is hybrid charging networks where context and urgency would dictate whether the vehicle satellites move between wireless and plug-in. Wireless charging lanes for public buses, logistics trucks, and airport shuttles are already being tested in Europe and Asia, and preliminary findings on these trials suggest a strong operational advantage.

To summarize, the wireless charging technology is moving toward frictionless, automated, and integrated EV charging ecosystems. Economic feasibility, efficiency, and regulatory harmonization are the key challenges; however, converging wireless power transfer with digital infrastructure, smart grids, and autonomous mobility is the advised means into the future of transportation. Transitioning the industry from conventional to intelligent EV charging systems, wireless systems will go from a luxury to a necessary step toward future-proofed mobility.

16.3 ADVANCED DRIVER-ASSISTANCE SYSTEMS

Advanced driver-assistance systems (ADAS) are reshaping modern automotive engineering by the application of intelligent technology to vehicles for enhanced safety, control, and automations. By implementing autonomous mobility, ADAS undergird the removal of driver error, reducing accidents and improving the driving environment. ADAS is thus not a single technology from a systems engineering point of view. Rather, it consists of a variety of integrated features based on sensor fusion, real-time data processing, artificial intelligence (AI), and control systems. Such features assist with human driving by giving warnings, taking partial control over vehicle functions, or even executing driving tasks under certain conditions. By adding ADAS to EVs, a layer of digital intelligence is integrated into transportation, increasing not only their sustainability but also their safety and operational efficiency. This section discusses the technical underpinning, core components, classifications, implementation approaches, performance indicators, deployment characteristics, and pathway to full autonomy.

Conceptual Foundation and System Architecture

At its heart, the ADAS intends to monitor the driving environment, interpret dynamic road conditions, and assist the driver in making a safe decision or intervene when necessary. A layered architecture with a focus on perception, decision, and actuation is the foundation of ADAS. The sensors of the perception layer comprise cameras, radar, LiDAR, ultrasonic sensors, and GPS modules, which register data about the surrounding vehicles, pedestrians, lane markings, traffic signs, and state of the road. The decision layer consists of algorithms that interpret the sensor data, using either machine learning or rule-based systems. It is here that the decisions are made regarding whether to alert or intervene with the vehicle. Lastly, the actuation layer carries the commands from the decision-making layer to the vehicle systems, such as braking, steering, and acceleration, to perform the required corrective intervention.

The control unit of the ADAS shall be considered to act as the central processor that takes in data from all sensors, runs software-based logic, and provides input to the vehicle's electronic control units (ECUs). As complexity increases, vehicles are transiting to domain architectures where the functions of ADAS are consolidated into central computing platforms, improving latency and coordination between such systems. Communication among these components must be both robust and of low-latency, using controller area network (CAN), automotive Ethernet, and dedicated communication buses. Table 16.11 provides an overview of the major components of the ADAS and their functional responsibilities.

Each sensor type has its strengths and limitations. For example, cameras can pick up colors and symbols, but their performance drops significantly under low-light rainy or foggy conditions. Radar does well even under poor weather conditions, although it does not have a good resolution. LiDAR has an awesome resolution and provides 3D mapping, but it is costly and tends to be sensitive to dirt or to interference from the environment. Hence, full situational awareness can only be realized through the fusion of many sensors.

Classification of ADAS Features by Level of Automation

ADAS functions can be classified along levels of driving automation supported by the SAE according to the J3016 standard. From this classification,

Table 16.11 Core Components of an ADAS Architecture.

Component	*Function*	*Technology Basis*
Camera systems	Lane detection, traffic signs, pedestrian recognition	Complementary metal-oxide-semiconductor (CMOS) sensors, computer vision
Radar sensors	Distance measurement to moving objects	Short- and long-range radar
LiDAR sensors	3D mapping and object contour detection	Light detection and ranging
Ultrasonic sensors	Parking assistance, low-speed maneuvers	Ultrasonic frequency emitters
ADAS ECU	Data processing and decision-making	High-performance SoCs
Connectivity modules	V2X communication, cloud data exchange	5G, DSRC, GPS, Wi-Fi

there are five levels ranging from Level 0, meaning there is no automation, through Level 1 and Level 2 systems, which are the ones that almost all trade vehicles come with at present, to Levels 3 and beyond, which are in trial testing in high-end and pilot platforms. Table 16.12 shows the classification of ADAS features based on the level of automation and system responsibility.

Such an expansion of Level 1 and Level 2 systems in commercial vehicles has been the result of regulation incentives and also a clear consumer demand for safer and partially autonomous driving. Levels 3 and beyond will require enormous advancements in perception accuracy, regulatory approval, and fail-operational system design.

Key Functional Domains of ADAS

ADAS define a wide variety of functions, which can be classified into several so-called core domains: longitudinal control [19,20], lateral control, parking assistance, visibility enhancement, and driver monitoring. Longitudinal control means controlling the acceleration and deceleration based on surrounding traffic, mainly using adaptive cruise control or forward collision mitigation. Lateral control means maintaining lane position by lane-keeping systems. Parking assistance can detect obstacles using ultrasonic sensors and cameras to plan a trajectory. Visibility enhancement includes automatic high beams, night vision, and rain-sensing wipers. Driver monitoring systems keep the driver vigilant and awake, often using technologies like tracking where the eyes are pointed or facial recognition identification.

Every domain requires some particular sensors and software modules. Coordination between the domains is very important, especially in complicated driving scenarios where simultaneous inputs are needed. Table 16.13

Table 16.12 ADAS Features by SAE Automation Levels.

SAE Level	*Description*	*Driver Role*	*Representative ADAS Features*
Level 0	No automation	Full control	Collision warning, lane departure alert
Level 1	Driver assistance	Control with support	Adaptive cruise control, lane-keeping assist
Level 2	Partial automation	Monitors environment	Highway assist, traffic jam assist
Level 3	Conditional automation	Human takeover upon request	Automated lane change, autonomous valet
Level 4	High automation	No driver needed in geofenced areas	Urban autonomous shuttle
Level 5	Full automation	No driver under any condition	Fully driverless taxis

Table 16.13 Functional Domains and Enabling Technologies in ADAS.

ADAS Function	*Operational Domain*	*Primary Enabling Technologies*
Adaptive cruise control	Longitudinal control	Radar, camera, real-time decision engine
Lane departure warning	Lateral control	Camera, neural network for line detection
Automatic emergency braking	Collision avoidance	Radar, LiDAR, brake-by-wire systems
Parking assist	Low-speed maneuvering	Ultrasonic sensors, 360° cameras
Driver drowsiness detection	Driver monitoring	IR camera, facial landmark tracking
Night vision enhancement	Visibility support	IR camera, thermal imaging sensors

gives a detailed overview of the common ADAS functions mapped into the operational domains and enabling technologies.

Key features of ADAS include real-time data processing, V2I communication, and automation alerts, which are designed to function despite environmental factors like weather. For instance, snow or fog can block cameras and LiDARs, which leads to reduced system reliability even under these adverse conditions. Whenever a sensor is impaired or fails, others can compensate due to their fusion and redundancy strategies.

Safety validation of ADAS poses serious challenges. Conventional testing methods for vehicles are inadequate to cover the vast number of edge cases that an ADAS-equipped vehicle may encounter. Hence, it must rely on simulation environments, digital twins, and a huge amount of real-world driving data for full coverage in testing. Ethical issues, such as how to encode collision-avoidance logic for moral edge cases, remain unresolved.

This isn't the whole picture. Table 16.14 provides the advantages and disadvantages of the major ADAS technologies along environmental and operational dimensions.

These limitations point toward the need for diversity of sensors and strong software algorithms to work with imperfect input data. High-reliability ADAS platforms typically employ techniques for fault detection and self-diagnosis to ensure plausibly safe operation under uncertainty.

Trends in Industry, Deployment, and Market Penetration

The last decade has witnessed the rapid integration of ADAS in mainstream vehicles. Manufacturers are typically forced to include core functions in standard equipment through a raft of regulation from such country-specific

Table 16.14 Environmental Sensitivity and Functional Limitations of ADAS Sensors.

Sensor Type	*Strengths*	*Limitations*	*Environmental Sensitivity*
Radar	Works in all weather, long-range	Low resolution, poor object classification	Low
Camera	High detail, object recognition	Affected by glare, fog, low light	High
LiDAR	Accurate 3D mapping	Expensive, dust-sensitive	Medium
Ultrasonic	Precise at short range	Ineffective beyond 3–5 meters	Medium
Infrared camera	Night vision capability	Limited detail, costly	Low

agencies as the European Union's General Safety Regulation and the US National Highway Traffic Safety Administration (NHTSA). As consumers outside were getting increasingly aware and demanding about safety and convenience, advanced features also quickly found their way into the mid-range segment of vehicles.

The development of EVs further gives impetus to the ADAS deployment, as both technologies have a common reliance on software-defined architecture and advanced electronics. Automakers are increasingly moving toward centralized domain controllers that support over-the-air (OTA) software updates, allowing continuous improvement and expansion of features after sales. Table 16.15 shows the comparison of the ADAS rates of adoption across different major automotive markets.

The localized differences in adoption are due to the economic systems that regulate their functioning, along with the infrastructure to support uses. The developed markets that lead to consumer protection and hold purchasing power are outgrowing the practical and economical use of entry-level ADAS functions that emerging economies are beginning to adopt due to popular availability of cost-effective hardware.

Future Outlook and Convergence on Autonomous Driving

ADAS is the foundation on which autonomous vehicle capabilities are laid. Gradually, high-definition mapping, 5G vehicle-to-everything (V2X) communication, and AI will enable higher levels of autonomy. Combining ADAS and autonomous driving systems enhances perception accuracy and machine learning, ensuring safety through system redundancies.

Table 16.15 Global ADAS Adoption Trends by Region.

Region	*Primary ADAS Features Deployed*	*Adoption Rate (2024 est.)*	*Regulatory Mandates (Yes/No)*
North America	ACC, AEB, lane assist	~78% (new vehicles)	Yes
Europe	ACC, AEB, ISA, parking assist	~85%	Yes
Asia-Pacific	ACC, lane watch, driver monitoring	~65%	Varies by country
Middle East	Blind spot monitoring, lane alert	~40%	No
Latin America	Basic alert systems	~25%	No

The birth of shared autonomous vehicles, robotic delivery platforms, and intelligent traffic systems will promote the creation of a symbiotic ecosystem where ADAS-equipped vehicles will communicate not only with human beings but also with infrastructure and machines. Learning will take place on the cloud, enriched with edge computing and real-time analytics.

In conclusion, as a key step toward achieving complete vehicle autonomy, ADAS provides intelligence to vehicles, improves safety by reducing accidents, and offers a more pleasant driving experience. Sensor limitations, software validation, and regulatory alignment may create challenges along this path to ADAS integration; however, all these issues are accelerating the adoption of ADAS integration. Therefore, ADAS will become an architectural standard for intelligent, interconnected, automated transportation systems, as it extends to more vehicle segments.

REFERENCES

1. Manzini, A., Martinez García, L., & Harrivaara, P. (2025). Adoption of next-generation batteries: A survey-based analysis of user and expert perspectives. *Mineral Economics*. https://doi.org/10.1007/s13563-025-00492-z
2. Gallegos, M. V., Ángeles Hernández Fenollosa, M., Damonte, L. C., Villabrille, P., & Caregnato, P. (2025). ZnO obtained from spent alkaline batteries as photocatalyst for carbamazepine degradation. *Emergent Materials*. https://doi.org/10.1007/s42247-025-00995-z
3. Liu, H., Xu, Q., Liu, H., Jin, Y., Shao, X., Zhang, X., Lv, S., Zhang, Z., & Ma, C. (2025). P doping adjusts the coordination structure of single atoms to accelerate the catalytic bidirectional redox reaction in Li–S batteries. *Rare Metals*, 44(8), 5428–5438. https://doi.org/10.1007/s12598-025-03233-x
4. Li, M., Liu, X., Shi, C., Wang, J., & Yao, S. (2025). Preparation of Sn/Zn/Co-NPC composite anode materials via impregnation method for sodium-ion batteries with enhanced cycling stability. *Ionics*, 31(7), 6963–6974. https://doi.org/10.1007/s11581-025-06352-8

5. Liu, F., Liu, J., Lin, Y., Huang, X., Lei, S., & Li, Y. (2025). Formal recycling for retired electric vehicle batteries: Behavioral insights into consumer preferences in Chongqing. *Environment, Development and Sustainability*. https://doi.org/10.1007/s10668-025-06287-0
6. Das, S. R., Mohapatra, S. K., Kamila, N. K., Mohanty, S. P., & Swain, K. P. (2025). Smart urban transportation: Advancing IoT-based thermal management for lithium-ion EV battery packs. *Discover Internet of Things*, 5(1). https://doi.org/10.1007/s43926-025-00121-9
7. Zhang, K., Xu, S., Xu, R., Yu, T., Sun, Z., & Li, F. (2025). Constructing fast ion transport and stable interface in PEO-based solid-state lithium metal batteries with bifunctional additives. *Science China Chemistry*. https://doi.org/10.1007/s11426-025-2625-3
8. Zeng, F., Jin, J., Liu, S., Sheng, G., Schwingenschlögl, U., Palisaitis, J., Zhou, J., Rosen, J., & Tu, S. (2025). Two-dimensional Mo4/3B2–xTz boridene cathodes for Al-ion batteries. *Communications Materials*, 6(1). https://doi.org/10.1038/s43246-025-00852-w
9. Demirkan, M. T., & Dogrusoz, M. (2025). Electrochemical reactivity of PZT materials in Li-ion and Na-ion batteries. *Discover Electrochemistry*, 2(1). https://doi.org/10.1007/s44373-025-00020-2
10. Saliu, O. D., Iresemowo, O., Kubi, F., Moberuagba, K. H., Adeniyi, A. G., & Ramontja, J. (2025). Enhancing the performance of MoO3-based cathodes for aqueous zinc-ion batteries: Spinel CuMoO3 vs non-spinel CuO–MoO3. *Journal of Materials Science: Materials in Engineering*, 20(1). https://doi.org/10.1186/s40712-025-00292-x
11. Jain, V. K., Kumari, A., Singh, S., & Tyagi, V. (2025). Electric vehicles as a pathway to decarbonisation in India: Assessing the environmental benefits and challenges of EV adoption for sustainable tomorrow. *Circular Economy and Sustainability*. https://doi.org/10.1007/s43615-025-00619-y
12. Wei, L. (2025). How to facilitate the recycling of retired electric vehicle batteries in China? A policy simulation based on system dynamics. *Journal of Material Cycles and Waste Management*. https://doi.org/10.1007/s10163-025-02330-4
13. Nishanth, G., Krishnan, M. M., Parandhaman, B., & Harinarayanan, J. (2025). Hardware implementation of EKF based SOC estimate for lithium-ion batteries in electric vehicle applications. *Scientific Reports*, 15(1). https://doi.org/10.1038/s41598-025-99931-8
14. Rajendran, S., Prakash, C., Shakir, M., Alwetaishi, M., Dhairiyasamy, R., Rajendran, P., & Lee, I. E. (2025). Enhancing competitiveness in India's electric vehicle industry: Impact of advanced manufacturing technologies and workforce development. *Scientific Reports*, 15(1). https://doi.org/10.1038/s41598-025-97679-9
15. Liu, B., & Han, J. (2025). Assessment of recovery potential and economic benefits from end-of-life new energy vehicles batteries in China based on GRA-BiLSTM. *Journal of Sustainable Metallurgy*, 11(2), 1381–1403. https://doi.org/10.1007/s40831-025-01049-8
16. Hong, G., & Choi, W. (2025). Utilizing ensemble models under various discharge conditions for static capacity estimation to promote the reuse of retired EV batteries. *International Journal of Precision Engineering and Manufacturing-Green Technology*. https://doi.org/10.1007/s40684-025-00701-8

17. Mirizzi, L., Carena, E., Santoro, C., Ficca, V. C. A., Placidi, E., Berretti, E., Lavacchi, A., Ferrara, C., & Muhyuddin, M. (2025). Many lives of cobalt within electrochemical systems: From waste batteries to efficient hydrogen evolution reaction electrocatalyst. *Materials for Renewable and Sustainable Energy*, 14(2). https://doi.org/10.1007/s40243-025-00312-y
18. Abdulrahman, G. A. Q., & Qasem, N. A. A. (2025). Assessment, optimization, and utilization mapping of clean and renewable energy technologies for powering unmanned aerial vehicles. *Arabian Journal for Science and Engineering*. https://doi.org/10.1007/s13369-025-10138-5
19. García-Soriano, F. J., Cometto, F., Raviolo, S., Slosar, T., Tchernychova, E., Genorio, B., Dominko, R., Bracamonte, M. V., & Vizintin, A. (2025). Biocarbon from olive pomace residue as a sulfur host for carbonate-based lithium-sulfur batteries. *Communications Materials*, 6(1). https://doi.org/10.1038/s43246-025-00846-8
20. Samiei Moghaddam, M., Azadikhouy, M., Salehi, N., & Hosseina, M. (2025). A multi-objective optimization framework for EV-integrated distribution grids using the hiking optimization algorithm. *Scientific Reports*, 15(1). https://doi.org/10.1038/s41598-025-97271-1

Chapter 17

Industry Collaboration and Partnerships

17.1 COLLABORATIONS BETWEEN AUTOMAKERS AND TECH COMPANIES

The global automobile industry is experiencing unprecedented convergence between traditional and modern technology companies. Since electric vehicles (EVs) are becoming more complicated, artificial intelligence (AI) is advancing rapidly, and connected, autonomous, and energy-efficient transport systems are emerging, this relationship is vital. As EVs become digital, these relationships redefine automotive engineering and digital innovation. This chapter discusses automaker–tech firm collaborations that promote electric transportation technologies.

Automakers separated operations because their in-house resources focused on mechanical engineering, manufacturing, and supply chain management. However, with the advent of electrification, automated driving, over-the-air updates, vehicle-to-everything (V2X) communication, and data-driven services, automakers have realized that the traditional approach to manufacturing was inadequate. Technology companies [1–3], with their strengths in software development, cloud computing, AI, and machine learning, have emerged as indispensable allies in enabling the software-defined vehicle paradigm. The project rationale extends far beyond technology integration, encompassing joint R&D, risk-sharing, cost optimization, intellectual property exchange, and faster time-to-market.

An increasingly important driving force in such collaboration is the increasing volume of demand for ADAS and fully autonomous driving platform solutions. Developing safe and reliable autonomous systems demands high-definition mapping, sensor fusion algorithms, real-time data processing, and large computing capacity—not normally found within de facto suppliers of legacy automakers. Thus, collaborative agreements between car manufacturers and AI companies have been on the rise. The relationship can further be illustrated through the GM–Cruise–Microsoft partnership, in which GM brings in the automotive platforms, Cruise develops the new autonomous technology, and, finally, Microsoft adds value in terms of leveraging its Azure cloud infrastructure for scalability.

 DOI: 10.1201/9781003608097-17

Battery technology and energy management are two other areas in which active collaboration is seen. Although traditionally automakers managed the integration of batteries and tuning of their performance, tech companies today contribute advances in battery management systems, AI optimization for performance, and data analytics. Northern Americas Partnership—formed by both Tesla and Panasonic—revolves around battery cell production. However, it does display the way in which software analytics can be coupled with chemistry to bring out a better range, longevity, and safety. Likewise, other regions under discussion would include companies such as CATL and LG Energy Solution, which partner with various global automakers on high-energy-density battery modules and solid-state battery prototypes.

From here, one can establish cross-cutting partnerships based on these various categories, enabling a view of the technology dimension at which partnerships enter to solve different problems.

Interestingly, the diversified nature of partnerships in the industry are shown in the following. In autonomous driving, deep AI and sensor fusion are central to collaboration with Nvidia and Waymo. Infotainment systems are now being developed more in cooperation with Google and Apple for seamless digital experiences for drivers and passengers, and these systems are well-aligned with consumer expectations of smart devices and cloud-connected ecosystems.

A key subtheme in this regard is how software-defined vehicles are developing. Unlike static firmware automobiles, EVs need software updates, diagnostics, cybersecurity, and data analytics. Tesla pioneered over-the-air (OTA) updates, but cloud service companies are now following suit. Volkswagen Group and Microsoft Azure Automotive Cloud collaborate to link vehicles for upgrades, real-time data analytics, and predictive maintenance.

Table 17.1 Classification of Automotive–Tech Collaborations by Integration Domain.

Domain of Collaboration	*Primary Automotive Partner(s)*	*Primary Tech Partner(s)*	*Key Objective*
Autonomous driving	General Motors, Toyota	Cruise, Waymo, Nvidia	Develop and deploy self-driving technologies
Battery management	Tesla, Volkswagen	Panasonic, CATL	Optimize battery performance and life cycle
Cloud and data services	Ford, BMW	Amazon AWS, Microsoft	Enable connected services and vehicle telematics
In-vehicle infotainment	Mercedes-Benz, Hyundai	Google, Apple	Integrate smartphone ecosystems with car systems
Mobility and ride-sharing	Honda, Toyota	Uber, Didi, Lyft	Launch EV-based ride-hailing platforms

These advances offer a new way to evaluate product integration and value generation collaborations. Table 17.2 illustrates how partnerships create value across the EV product lifecycles.

Table 17.2 offers a clear view of how partnerships cross-cut the vehicle lifecycle. During the design phase, digital twin technologies allow real-time simulations of vehicle behavior for several scenarios. An example is BMW working with Nvidia to use the Omniverse platform for the simulation of complex environments and virtual vehicle behavior, drastically shrinking the time required for physical testing. Google Cloud then provided cloud-based analytics tools used by Ford for monitoring assembly lines, tracking production defects, and enabling AI-based decision-making for quality control purposes.

Alliances affect consumer experience. Google and Hyundai have integrated voice assistants, navigation, and other streaming services into vehicle dashboards. This connectivity makes electric mobility more comfortable and convenient for digital natives who want their cars to be mobile smart devices and deployments.

IP exchange and platform standards are key to these partnerships. Automakers may use an open-source software stack and shared standards as vehicles become software-controlled. Volvo, Polestar, and GM fitted android automotive operating system (AAOS). A unified software environment streamlined app, interface, and model updates.

Co-ownership and equity investments connect long-term interests in partnerships. GM invested in Cruise and Amazon in Rivian, which showed financial and operational integration. The investments give IT businesses and automakers early access to the investors' cutting-edge software and AI.

Collaborations need huge effort and commitment. Data on high-value partnerships with financial estimates and targets are presented in Table 17.3.

Table 17.2 Value Creation through Automotive–Tech Partnerships.

Stage of Product Lifecycle	*Collaborative Contribution*	*Example Partnership*	*Value Delivered*
Design and development	Shared R&D platforms, digital twins	BMW and Nvidia	Accelerated vehicle prototyping
Production and testing	Smart factory and automation tools	Ford and Google Cloud	Optimized manufacturing processes
Deployment and launch	OTA readiness, data connectivity	VW and Microsoft Azure	Scalable software deployment
Customer use phase	AI-based diagnostics, infotainment	Hyundai and Google Assistant	Enhanced user experience and engagement
Post-sale and support	Predictive maintenance, analytics	Tesla and internal AI stack	Reduced downtime and operational cost

Table 17.3 High-Value Automotive–Tech Collaborations.

Collaboration Partners	*Investment Size (USD)*	*Year Initiated*	*Primary Objective*
GM–Cruise–Microsoft	$2 billion	2021	Commercialization of autonomous EVs
VW–Microsoft	$1 billion	2019	Build automotive cloud for VW Group
Ford–Google	$900 million	2021	Data analytics and AI integration
Amazon–Rivian	$700 million	2019	Electrification of delivery fleet
Volvo–Nvidia	Undisclosed	2020	Develop AI-based vehicle platforms

With investments topping hundreds of millions or billions, this table indicates how important automotive–tech alliances are. Each cooperation addresses AI, cloud infrastructure, electrified logistics, and other gaps. Rivian and Amazon are collaborating to create electric delivery trucks and scalable logistics infrastructure for Amazon's e-commerce operations.

Geographical aspects should also be highlighted. Although the United States tends to remain the operational core for many of these high-profile partnerships, Europe and Asia have become other active centers. Chinese automakers BYD and NIO are partnering with domestic and global tech firms for intelligent driving platform [4–6] development and battery innovation boosting. Moreover, these partnerships are being influenced by the regional regulatory environment, data privacy laws, and government incentives for local manufacturing and R&D.

To capture the cross-sectoral nature of these partnerships, Table 17.4 maps out the flow of expertise from tech companies to automakers and vice versa.

Table 17.4 shows that the knowledge flows are not unidirectional in nature. While it is true tech companies are poising themselves with innovations in AI and data analytics, the automakers bring in deep domain knowledge with respect to vehicle systems engineering, crashworthiness, and compliance. Hybrid companies can build robust, digital cars through joint ventures and co-development laboratories.

Automakers and technology businesses must work together to thrive in the electric and intelligent mobility revolution. They aid electrification, energy storage, vehicle autonomy, user experience, and sustainability. These industries will collaborate to rethink transportation through cooperative inventions, complementing skills, and shared knowledge in these innovative times.

Table 17.4 Knowledge and Capability Flow in EV Collaborations.

Flow Direction	*From (Sector)*	*To (Sector)*	*Type of Capability Transferred*
Technology to auto	Cloud, AI, UX firms	Automakers	Machine learning, user experience design
Auto to technology	OEMs, Tier 1s	Tech companies	Vehicle engineering, safety testing
Bidirectional	Joint ventures	Joint ventures	Co-developed software stacks, analytics

17.2 ROLE OF STARTUPS AND INNOVATION HUBS

OEMs and Tier 1 suppliers dominated electrified cars, with some changes. They challenge the "establishment" in thinking and applications by deploying disruptive technology in their value chains that conventional players may be slow to accept. Accelerators, academic research parks, and government-sponsored tech clusters develop, test, and scale mobility technologies for commercialization.

Startups and innovation hubs play multiple roles in EVs. Startups typically specialize on battery chemistry, fast-charging algorithms, thermal management, autonomous navigation, and blockchain-based EV energy trading. Deep domain expertise and rapid prototyping enable these directed efforts' technical work, which increasingly uses venture finance and academic partnership. Innovation hubs reduce market entry hurdles for innovative ideas with infrastructure, mentorship, regulatory assistance, and pilot projects. Innovation becomes broad, experimental, and data-driven using these formulas.

Startups add maximum value to EVs through battery innovation. Battery producers have long production cycles and risk-averse product development. To reduce battery deterioration, range anxiety, safety, and energy density, QuantumScape, StoreDot, and Solid Power pursue solid-state electrolytes, ultra-fast-charging materials, and silicon-dominant anodes. High-risk, high-reward models in research and development allow startups to pursue materials and architectures that would be hard to justify without early proof-of-feasibility in large-scale manufacturing environments.

This establishes better knowledge [7–9] of the areas of startup contributions. Table 17.5 provides the classification of startup activity areas across key EV technology domains.

Table 17.5 outlines the different and complementary ways startups contribute to the ecosystem. For example, the gallium nitride (GaN)-based power electronics from companies such as Navitas provide smaller, more efficient designs of inverters for EV powertrains, resulting in reduced energy losses and lighter vehicle designs. Meanwhile, software-based startups like

Table 17.5 Technological Focus Areas of EV Startups.

Focus Area	*Representative Startups*	*Core Innovation*	*Industry Impact*
Battery and energy storage	QuantumScape, Solid Power	Solid-state batteries, fast charging	Increased energy density and safety
Charging infrastructure	EVBox, ChargePoint, FreeWire	Smart chargers, mobile charging units	Expanded access to fast and mobile charging options
Power electronics	Navitas, Pre-Switch	GaN-based inverters, power management integrated circuits (ICs)	Enhanced efficiency and thermal performance
Autonomous software	Ghost Autonomy, Oxbotica	Modular autonomy stacks	Lower-cost, scalable self-driving platforms
Fleet and data management	Revel, ViriCiti	Telematics, battery diagnostics	Improved uptime and operational control

Ghost Autonomy furnish autonomous driving stacks that can be integrated into both passenger and commercial EVs, thus decreasing reliance on vertically integrated OEM efforts.

With startup growth, innovation hubs can help ecosystem collaboration. Test tracks, simulation labs, prototype labs, and regulatory sandboxes enable experimentation. Government, university, and private-sector collaborations support innovation hubs to reduce early-stage business risk. Such hubs can mix software engineering, material science, electrical engineering, and behavioral psychology to develop mobility-centric products.

Innovation hub architecture varies globally. While some are globally networked, localized innovation hubs focus on urban mobility in heavily populated areas. Table 17.6 describes innovation hub structures. Innovation centers support multiple phases and innovation sets, as shown in this table. University hubs conduct foundational research and prototype creation, necessitating academic competence and elite labs. Unlike university-affiliated hubs, corporate accelerators like BMW Startup Garage focus on fast-to-pilot ideas. Federal investment in public infrastructure and pilot deployment environments supports state development goals like pollution reduction, job growth, and transportation equity.

Flexible innovation network nodes are another benefit of new ventures. Unlike traditional manufacturers, startups may "plug-and-play" their products on OEM and cloud platforms. This supports open innovation frameworks, where stakeholders share knowledge and resources to co-create solutions. Thus, they fare well in erasing bottlenecks to innovation [10–12], speeding time-to-market, and causing cross-pollination of ideas across sectors.

Table 17.6 Typology of Innovation Hubs Supporting EV Startups.

Innovation Hub Type	*Organizational Model*	*Support Services Offered*	*Examples*
University-affiliated	Academic–government partnership	Research labs, faculty mentorship, student talent	MIT Mobility Lab, TU Delft Incubator
Corporate accelerators	OEM or Tier 1 sponsored	Funding, product testing, joint R&D	BMW Startup Garage, Ford X
Government-funded clusters	Regional development agencies	Subsidies, pilot programs, infrastructure	Silicon Valley Clean Tech Cluster, ARAI Hub
Independent incubators	Private venture capital backed	Seed capital, networking, technical mentorship	Plug and Play Mobility, Y Combinator
International innovation zones	Multination collaborative	Trade facilitation, standards alignment	European Battery Alliance, ASEAN EV Hub

One of the starting points to appreciate this value that comes through these networks is by analyzing the trends in startup investments. In this funding category, we see most such developments with the much-needed venture capital scaling vast amounts of early-stage technologies to commercial readiness. Table 17.7 presents EV-related startup funding over the past five years, segmented by domains.

On funding, however, battery innovation and autonomous driving have attracted the highest investments, but charging infrastructure and vehicle software systems are catching up rapidly. These investments reflect investor confidence in enterprises overcoming systemic concerns including energy storage reliability, charge point accessibility, and software modularity.

Commercialization pipeline placements are the best sector-wide markers of startup success. EV firms often co-develop components with OEMs and become acquisition targets or suppliers. This link helps larger firms share knowledge and reduce risk. Cruise, a company turned GM autonomous driving subsidiary, is one example. Arrival, Rivian, and Proterra began manufacturing due to close relationships.

Geographical clustering aids startups. Tech-centric innovation centers provide talent, venture financing, regulatory cooperation, and advanced infrastructure. Table 17.8 provides a comparison of major global EV startup clusters.

Table 17.8 shows that regions specialize in EV development based on government frameworks, industrial legacy, and population demands. East Asia leads in battery and sensor technologies due to manufacturing and government support, whereas Europe stresses sustainability and legislative harmonization. North America stresses on software ecosystems, capitalization, autonomy, and logistics platforms. India might lead grid-compatible charging and frugal innovation.

Table 17.7 Venture Capital Funding in EV Startups by Sector (2019–2023).

Domain	*Total Funding (USD Billion)*	*Top Funding Year*	*Notable Funded Startups*	*Growth Rate (CAGR %)*
Battery innovation	$9.3	2022	Northvolt, Sila Nanotechnologies	17.2
Charging infrastructure	$6.8	2021	ChargePoint, Volta	15.4
Autonomous driving	$10.1	2020	Aurora, Pony.ai	13.7
Fleet and logistics tech	$4.2	2023	Revel, Geotab	11.1
Vehicle software systems	$5.5	2022	Sibros, Sonatus	14.3

Table 17.8 Leading Regional Hubs for EV Startup Activity.

Region	*Key Cities*	*Specialization Areas*	*Supporting Institutions*
North America	San Francisco, Austin, Detroit	Battery tech, autonomy, fleet software	DOE Labs, ARPA-E, Techstars Mobility
Europe	Berlin, Stockholm, Eindhoven	Sustainable materials, e-mobility	EIT InnoEnergy, Fraunhofer Institutes
East Asia	Shenzhen, Seoul, Tokyo	Battery manufacturing, sensors	Tsinghua University, KAIST, METI Japan
India	Bengaluru, Pune, Hyderabad	Affordable EVs, charging infra	NITI Aayog, ARAI, iCreate
Middle East	Dubai, Abu Dhabi	Smart mobility platforms	Masdar Institute, Hub71

Finally, start-ups and innovation hubs boost technological innovation within the EV ecosystem. They challenge incumbents, democratize innovation, and encourage collaboration on transdisciplinary concerns like climate change, urban congestion, and energy constraint. A changing industrial landscape requires speed, risk-taking, and constant learning, which assist in developing electric mobility.

Startup integration into mainstream development pipelines may determine the future of EVs. Table 17.9 shows that startups and established automakers have various cooperative innovation ecosystem advantages.

Notably, given the clear picture it paints, this table emphasizes that innovation is at its optimum when both types of organizations work within structured frameworks: “start-ups and traditional OEMs.” The latter

Table 17.9 Comparative Innovation Capabilities of Startups Versus Traditional OEMs.

Innovation Factor	*Startups*	*Traditional Automakers*	*Synergistic Outcome*
Speed of prototyping	High	Moderate	Rapid concept validation
Access to capital	Risk capital, agile rounds	Stable corporate finance	Blended financial resilience
Regulatory compliance	Limited	Extensive	Guided regulatory navigation
Manufacturing capability	Limited	Extensive	Scalable production of validated ideas
Market credibility	Emerging	Established	Accelerated trust-building and adoption

provides scalability, reliability, and regulatory compliance, while the former directs experimentation and pace. Accelerators and innovation centers combine skills.

In conclusion, electric car startups and innovation centers are growing in importance. They bridge technological gaps and change innovation culture. They will produce competitive advantage engines as the global EV ecosystem matures. These groups promote decentralized, collaborative, fast industrial innovation. Successful ecosystems provide capital, infrastructure, and mainstream market integration. Technology and innovation architecture that integrates startup brilliance with traditional industry leaders' networks will shape electric transportation's future.

17.3 PUBLIC–PRIVATE PARTNERSHIPS

Public–private partnerships (PPPs) may be the most essential tool for global electric car development, deployment, and adoption. Sharing resources, planning, and risk are possible with PPPs. Businesses and governments collaborate on electric transportation PPPs to address environmental goals, market failures, infrastructure constraints, and policy uncertainties. PPPs may scale up innovation, enablement, and inclusive and equitable access to electric mobility as the global migration of internal combustion engines to zero-emission alternatives intensifies.

The EV ecosystem is especially supportive of PPPs because launching an EV requires integrating almost all systems [13–15]: vehicle construction, charging infrastructure, energy grid compatibility, regulatory foundations, recycling systems, and consumer end-of-life behavior. No one group of stakeholders has the technical expertise, jurisdictional authority, and financial resources to handle these elaborate requirements. Governments establish PPPs to gain access to the private sector's innovations, supply chain logistics, and capital investment. In return, private firms benefit from regulatory certainty or market incentives and often receive access to public land or utilities and praise for collaboration with public institutions. These partnerships bridge, therefore, institutional and operational gaps among regulation, innovation, and implementation.

In the EV field, the structure of PPPs varies remarkably from one area of application, scope of involvement to governance model. Some PPPs have been formed to facilitate large infrastructural undertakings such as national networks for vehicle charging; others are focused primarily on innovation—joint research, pilot programs, or mobility labs of cities. The diverse character of these partnerships represents the aims on which Table 17.10 has been based—the typology of PPPs according to the operating objectives in the EV sector.

Table 17.10 illustrates some of the application areas for PPPs within the EV sector. The most visible infrastructure types of partnerships can be found in areas that have a stronger state position in energy systems. For example, fast-charging corridors on highways and in metropolitan areas are the objectives of national PPPs in the United States, India, and Germany, where the federal transportation department generally collaborates with utility providers and private EV supply equipment (EVSE) manufacturers. Other types of partnerships are focused on innovation, such as the European Battery Alliance, which brings together private industry actors with funding research mechanisms and technical institutes in order to address the need for competitive battery manufacturing capacity within the European Union.

Public–private cooperation is another key role of PPPs by mobilizing demand through fiscal and non-fiscal incentives. Especially where consumers are still adjusting to low upfront EV costs and awareness of electric

Table 17.10 Typology of PPPs in the EV Ecosystem.

PPP Type	*Primary Objective*	*Key Stakeholders Involved*	*Example Initiative*
Infrastructure development	Expansion of charging stations, grid access	Energy utilities, EVSE firms, transport agencies	National EV Charging Initiative (United States)
Policy and incentive design	Joint development of subsidies or mandates	Regulatory bodies, automotive associations	EV30@30 Campaign (IEA, UN, OEMs)
R&D and innovation	Joint research in battery or vehicle tech	Universities, research councils, OEMs	EU Battery Alliance
Urban mobility programs	Integration of EVs in public fleets	City governments, transit operators	London Electric Bus Partnership
Recycling and sustainability	Battery reuse, material recovery	Recycling firms, environmental agencies	Li-Cycle and New York State Collaboration

mobility is limited, this is particularly important. Such partnerships would jointly design and implement incentive schemes for tax credit and rebate policies, low-interest financing, and scrappage programs. The outreach to consumers and fleet operators would be further coordinated through the partnerships. With incentives being co-created for such systems, both manufacturers and users guarantee financial viability and availability. Table 17.11 presents a comparative analysis of public–private incentive schemes at the national level and their effect on the adoption and diffusion of EVs.

Table 17.11 illustrates the efficiency of the partnership between policy instruments, which is successful when executed in collaboration with private-sector stakeholders. Thus, in Norway, the partnership between automotive industry bodies and policymakers is aimed toward building comprehensive consumer incentives and infrastructure support [16, 17], leading to very high EV adoption worldwide. In this context, incentives for the purchase of an EV are to be shared between OEMs and the federal government, thereby allowing wider market access in Germany. On the other hand, major government interventions in India have been through PPPs, making possible near-rapid diffusion of electric two- and three-wheelers, especially from specific schemes under the FAME policy.

In addition to national-level programs, PPPs are also being employed by local governments to change the patterns of mobility in cities. Most municipalities have collaborated with private mobility providers, charging station

Table 17.11 Comparison of the Impact of Public–Private Incentive Programs on EV Adoption.

Country	*Type of PPP Incentive Program*	*Private Partner Role*	*EV Adoption Rate (as of 2023)*	*Notable Outcome*
Norway	Tax exemption, toll discounts	Auto associations promote compliance	>80% of new car sales	World's highest EV penetration
China	Purchase subsidies, industrial grants	OEMs align with regional targets	~25% of new car sales	Rapid domestic EV industry growth
United States	Federal tax credits, charging incentives	Utilities and automakers co-fund charging infrastructure	~8% of new car sales	Nationwide charging corridor underway
India	FAME II subsidies, battery standards	Two-wheeler OEMs scale affordable EVs	~6% of new vehicle sales	Significant growth in electric scooters
Germany	Co-funded purchase bonus programs	OEMs share cost of incentives	~20% of new car sales	OEM–government price-sharing for buyers

vendors, and city planners to provide electric buses, shared mobility fleets, and last-mile delivery solutions. They are among the most significant contributors to air pollution, traffic congestion, and noise in dense urban centers. Some of the early adoption examples include London, Amsterdam, and Los Angeles, which have all engaged in such multi-stakeholder partnerships. Table 17.12 provides some selected examples of municipal-level urban e-mobility PPPs [18, 19] sets.

For example, in Amsterdam, predictive analytics combined with university research is utilized for integrating urban energy demand properties into grid flexibility within the framework of initiatives such as Smart Charging. The BlueLA project in Los Angeles is the way that equity-focused PPPs will be able to ensure that underrepresented communities will have access to EVs through targeting price and placement strategies. Such urban PPPs will frequently serve as test beds for the scalability of national systems while helping define standards and operational blueprints.

Another crucial area where development of PPPs is taking place is the sustainable lifecycle of EVs with respect to reuse and recycling of batteries, as well as circular economy frameworks. With increasing volumes of EV batteries reaching the end-of-life, governments and industries need to co-develop safe disposal strategies for those batteries, possible recovery of materials, and second-life applications. These projects require logistics, regulatory alignment, and recycling technological breakthroughs. Table 17.13 highlights key public–private battery recycling and material recovery operations.

The Swedish Green Batteries Alliance takes full responsibility for environmental compliance from manufacture, use, and recovery, minimizing the dependency on lithium, cobalt, and nickel imports and supporting

Table 17.12 Examples of Urban-Level Public–Private EV Mobility Initiatives.

City	*Program Name*	*Scope of Partnership*	*Primary Goals Achieved*
London	Electric Bus Deployment	City government + private bus operators	Transition to fully electric public transport
Amsterdam	Smart Charging Amsterdam	Municipality + energy firms + universities	Citywide smart charging integration
Los Angeles	BlueLA Car Sharing	City + car-sharing operator + community organizations	Affordable EV access in low-income neighborhoods
Shenzhen	Public Transit Electrification	City + OEMs (BYD) + grid operators	100% electrified city bus fleet
Montreal	EV Taxi Program	Local government + taxi firms + Hydro-Quebec	EV transition of the private taxi sector

Table 17.13 Public–Private Collaborations in Battery Recycling.

Program Name	*Country*	*Lead Partners*	*Objective*	*Notable Feature*
Li-Cycle and NYSERDA Program	United States	Li-Cycle + New York State Agency	Establish urban lithium-ion recycling facility	Urban-centered recycling pilot
BatRE Initiative	Germany	BASF + German Federal Ministry	Closed-loop battery material recovery	Integration with EU battery passport
Green Batteries Alliance	Sweden	Northvolt + Vinnova	Eco-friendly battery lifecycle management	Focus on ethical sourcing and recycling
ReCell Center	United States	Argonne National Lab + Industry Partners	R&D on battery reconditioning and reuse	Federally backed research consortium
India Battery Waste Program	India	NITI Aayog + Battery OEMs + Pollution Boards	National battery recycling framework	Compliance-driven extended producer responsibility

sustainability and resource security. These projects are essential to the circular economy of EVs.

PPPs have also evolved with increased governance complexity, adopting frameworks containing data-sharing agreements, outcome-linked metrics, and performance-based financing models. Such complex arrangements require formal institutional structures that guarantee transparency, accountability, and responsiveness to the public. Performance indicators, open data standards, and interoperability protocols are increasingly being utilized by public agencies to evaluate the outcome of partnerships. Table 17.14 elaborates on the governance features found in matured EV PPPs.

The following governance mechanisms will allow partnerships between the public and private sectors to operate with flexibility without losing public accountability. Performance-based contracts, for instance, guarantee incentive structures on the private partner to achieve usage targets or emission reduction goals. Data acquired under open access will benefit municipal planners, but this will also allow for load balancing and forecasts on urban mobility.

PPPs are the backbone of the growth of the EV ecosystem. It lays the structural and financial foundations for rapid infrastructure deployment, co-designed incentive regimes, targeted R&D investments, and equity in urban mobility. A successful governance structure will pool resources, share risks, and scale innovations across multiple layers of the value chains—from battery recycling and smart grid integration to urban fleet deployment and

Table 17.14 Governance Features of Mature EV PPPs.

Governance Feature	*Description*	*Strategic Value*
Performance-based contracts	Outcome-tied funding and evaluation frameworks	Ensures delivery of measurable results
Public land/asset access	Easement for charging stations and public EV fleets	Reduces capital barriers for infrastructure expansion
Open data protocols	Real-time access to usage, load, and pricing data	Enhances transparency and demand planning
Multi-stakeholder councils	Regular forums for consultation and feedback	Builds trust and long-term alignment
Regulatory flexibility zones	Sandboxes for experimentation with new mobility models	Encourages innovation with lower compliance risk

nationwide consumer outreach. Alliances have far-reaching implications. All nations can decarbonize, green, smarten, and make inclusive their transport networks with PPPs.

REFERENCES

1. Mondal, S., & Suman, N. K. (2025). Challenges in embracing electric vehicles: Exploring the policy landscape and way forward. *Transportation in Developing Economies*, 11(2). https://doi.org/10.1007/s40890-025-00248-y
2. Shankar, R. (2024). An integrated approach for modelling barriers of EVsCI. *International Journal of System Assurance Engineering and Management*, 16(1), 356–376. https://doi.org/10.1007/s13198-024-02623-1
3. Sharma, S. K., George, S. A., Srivastava, P. R., Jabeen, F., & Lafci, C. (2024). Supply chain socially sustainability practices and their impact on supply chain performance: A study from the Indian automobile industry. *Annals of Operations Research*. https://doi.org/10.1007/s10479-024-05991-w
4. Senbet, L. W. (2025). Multinational enterprises and climate action: A low-income perspective with Africa focus. *Journal of International Business Studies*, 56(4), 481–490. https://doi.org/10.1057/s41267-024-00764-w
5. Arshad, M. W., & Lodi, S. (2025). Quantum computing in the automotive industry: Survey, challenges, and perspectives. *The Journal of Supercomputing*, 81(10). https://doi.org/10.1007/s11227-025-07573-4
6. Mohd Radzi, N. A., Saidi, N. A., Hassan, S., Ibrahim, M. S. N., & Lee, K. E. (2025). Exploring corporate social responsibility practices in the telecommunications, broadcasting and courier sectors: A comparative industry analysis. *Humanities and Social Sciences Communications*, 12(1). https://doi.org/10.1057/s41599-025-04668-4
7. Li, Y., Li, Q., & Shi, J. (2025). Effects of dual-credit policy on electric vehicle diffusion under subsidy transfer: A multi-stakeholder perspective. *Environment, Development and Sustainability*. https://doi.org/10.1007/s10668-024-05926-2

8. Prakhar, P., Jaiswal, R., Gupta, S., Ahmad, S. Z., Piccardi, P., & Santoro, G. (2025). How can we improve entrepreneurial dynamics in electric vehicle manufacturing for a sustainable future: Insights using a deep learning-based hybrid PLS-SEM-ANN approach. *International Entrepreneurship and Management Journal*, 21(1). https://doi.org/10.1007/s11365-025-01072-x
9. Janamala, V., Rani, P. S., Rani, K. R., & Swarnasri, K. (2025). Transformative impact of electrical engineering on society, education, academia, and industry: A brief review. *Electrical Engineering*. https://doi.org/10.1007/s00202-025-03066-y
10. Simonazzi, A. (2025). The bumpy road of the European automotive industry towards sustainable mobility. *Rendiconti Lincei. Scienze Fisiche e Naturali*, 36(2), 417–428. https://doi.org/10.1007/s12210-025-01314-8
11. Malfliet, A., & Blanpain, B. (2025). Sustainable metallurgical ecosystems: Key enablers and lessons learned. *Journal of Sustainable Metallurgy*. https://doi.org/10.1007/s40831-025-01147-7
12. Fantin Irudaya Raj, E., Appadurai, M., Darwin, S., & Chithambara Thanu, M. (2025). Emerging technologies in power systems and their contribution towards sustainability in developing countries. *Polytechnica*, 8(1). https://doi.org/10.1007/s41050-024-00053-3
13. Zhao, K., Sanmugam, A., & Aravindan, K. L. (2025). Driving sustainability in the green entrepreneurial intentions among SMEs in the electric vehicle component industry. *Discover Sustainability*, 6(1). https://doi.org/10.1007/s43621-024-00667-8
14. Purnell, P. J. (2025). Transdisciplinary research: How much is academia heeding the call to work more closely with societal stakeholders such as industry, government, and nonprofits? *Scientometrics*. https://doi.org/10.1007/s11192-025-05367-2
15. Dhairiyasamy, R., Gabiriel, D., Bunpheng, W., & Kit, C. C. (2024). A comprehensive analysis of India's electric vehicle battery supply chain: Barriers and solutions. *Discover Sustainability*, 5(1). https://doi.org/10.1007/s43621-024-00595-7
16. Mahboubi, A., Miyangaskary, M. K., Keivanpour, S., & Lamghari, A. (2025). The sustainable shared reverse logistics framework: A blueprint for EV battery recovery. *Circular Economy and Sustainability*. https://doi.org/10.1007/s43615-025-00623-2
17. Awasthi, C., & Devi, P. (2025). Synergizing progress: Policies and advancements at the energy-material nexus. *International Journal of Advances in Engineering Sciences and Applied Mathematics*. https://doi.org/10.1007/s12572-025-00406-w
18. Dhairiyasamy, R., & Gabiriel, D. (2025). Sustainable mobility in India: Advancing domestic production in the electric vehicle sector. *Discover Sustainability*, 6(1). https://doi.org/10.1007/s43621-025-00844-3
19. Rajendran, S., Prakash, C., Shakir, M., Alwetaishi, M., Dhairiyasamy, R., Rajendran, P., & Lee, I. E. (2025). Enhancing competitiveness in India's electric vehicle industry: Impact of advanced manufacturing technologies and workforce development. *Scientific Reports*, 15(1). https://doi.org/10.1038/s41598-025-97679-9

Chapter 18

Consumer Behavior and Market Dynamics

18.1 CONSUMER BEHAVIOR AND PREFERENCES

Consumer behavior and preferences are crucial to EV adoption. Technology, infrastructure, customer thinking, attitudes, values, and practical decision-making are needed to convert from internal combustion engine vehicles (ICEVs) to electric automobiles. Psychological, economic, and social factors influence electric car perception, assessment, and adoption. Awareness, pricing, performance expectations, environmental conscience, risk aversion, and lifestyle compatibility affect customer behavior. Empirical facts, theoretical perspectives, and thorough subsections are used to study these traits.

Awareness and Knowledge

Electric vehicle (EV) adoption depends on consumer education, which highlights EV availability, technology, benefits, constraints, and incentives. Studies imply that while EV awareness boosts EV buying intent, battery range, charging availability, and long-term maintenance misconceptions prevent switching. Research suggests that education, dealership training, and experience marketing like EV test drives can help buyers understand and overcome these myths.

Economic Considerations

Most car owners buy for economic reasons. Consumers weigh upfront expenses, TCO, fuel savings, maintenance, and resale. Although EVs cost more upfront, their lower running costs can offset this. Savings come from government subsidies, tax credits, and incentives.

Table 18.1 shows the drop in EV ownership expenses over time despite the higher initial cost. Tax rebates and lower fuel and maintenance costs reduce 5-year ownership costs.

DOI: 10.1201/9781003608097-18

Table 18.1 Comparative Cost Analysis of ICEVs versus EVs over a 5-Year Period.

Cost Component	*ICEV (USD)*	*EV (USD)*
Purchase price	30,000	40,000
Fuel/electricity	9,000	3,000
Maintenance	4,500	1,200
Tax incentives	0	−7,500
Total 5-year cost	**43,500**	**36,700**

Table 18.2 Influence of Environmental Concern on Purchase Intent (Survey of 1,000 Respondents).

Level of Concern	*Likely to Buy EV (%)*
High concern	72
Moderate concern	48
Low concern	21

Environmental Values

More individuals care about the environment and believe that EVs can reduce greenhouse gas emissions, air pollution, and fossil fuel dependence. Urban and young consumers value the environment more. Political views, cultural norms, and climate change beliefs alter this influence. Psychologically, EVs' green image boosts consumers' environmental responsibility.

The above data in Table 18.2 illustrates that environmental concern greatly drives EV adoption intent.

A comprehensive summary is presented in Table 18.3, which provides a review of consumer and market dynamics across existing EV models.

Technological Readiness and Trust

Comfort with new gadgets varies. Battery, charging, car durability, and software trust affect EV purchases. Long-term reliability, battery degradation, and unfamiliarity with regenerative braking or digital interfaces may cause hesitation. Early adopters are tech-savvy and willing to try new technologies, whereas mainstream buyers wait for reliability and infrastructure.

Table 18.4 shows generational EV trust differences and that confident younger consumers adopt more.

Lifestyle Compatibility

Lifestyle suitability of an electric car heavily influences decision-making. EVs are better for urbanites with home charging and for short commutes.

Table 18.3 Review of Consumer and Market Dynamics of Existing EV Models.

Ref	*Method*	*Main Objectives*	*Findings*	*Limitations*
[1]	Stochastic optimization with green certificate market modeling	Manage energy in microgrids with electric bicycle charging and green certificates	Integrated approach improves sustainability and market participation	High computational load; assumes market price stability
[2]	Multi-objective optimization (PSO, GA)	Optimal siting of EV charging stations in hybrid renewable systems	Coordinated placement reduces system losses and enhances voltage profile	Limited scalability to large urban networks
[3]	Multi-objective optimization using pareto-frontal analysis	Manage EV fleet energy in smart microgrids considering economics	Effective balance among cost, emissions, and user comfort	Assumes deterministic load and renewable generation profiles
[4]	Scenario-based policy analysis	Evaluate EVs as a tool for decarbonization in India	EV adoption significantly reduces carbon footprint	Limited by uncertain policy and infrastructure development
[5]	Experimental thermo-fluid analysis	Enhance battery cooling with magnetic nanofluids and fields	Magnetic synergy increases thermal dissipation efficiency	Scalability and economic viability are yet to be validated
[6]	Reinforcement learning (deep Q-network)	Optimize EV charging/discharging in V2B-PV systems	Mobility-aware reinforcement learning (RL) improves energy usage and cost savings	Requires detailed mobility prediction and RL training overhead
[7]	Robust demand response modeling	Improve grid reliability under EV and wind uncertainty	Proposed framework reduces imbalance and enhances reliability	Complexity increases with larger uncertainty scenarios
[8]	Mixed-integer nonlinear programming (MINLP)	Schedule wind and EV loads with demand response	Achieves cost-effective operation with high renewables	High computational time for large datasets

(Continued)

Table 18.3 (Continued)

Ref	*Method*	*Main Objectives*	*Findings*	*Limitations*
[9]	Econometric modeling (VAR, GARCH)	Analyze market risks in green energy investments in India	Forex and commodity markets significantly affect EV investments	Based on historical data; may not predict future dynamics accurately
[10]	Analytical modeling and simulation	Compare ideal energy usage across EV powertrains	Identifies drivetrain configurations with optimal efficiency	Does not incorporate real-world driving cycles or consumer behavior
[11]	Machine learning (ML)-based classification and regression	Predict EV penetration using various ML algorithms	Random forest outperformed others in estimation accuracy	Dataset imbalance affected minority class prediction
[12]	System dynamics modeling	Forecast consumer EV adoption and automobile market evolution	Behavioral insights improve market trend prediction	Simplified consumer decision factors; does not cover heterogeneous preferences
[13]	Agent-based modeling (spatial)	Estimate hydrogen vehicle market potential in Beijing	Identifies spatial zones suitable for hydrogen vehicle rollouts	Limited generalizability to non-urban or non-Chinese contexts
[14]	Panel regression analysis	Investigate EV market penetration in rising motorization countries	Income and urbanization drive EV penetration more than policy alone	Lacks micro-level behavioral data
[15]	Battery performance benchmarking	Evaluate modern EV batteries on efficiency and longevity	Identifies chemistry-specific trade-offs for future design	Real-world aging effects not fully captured in lab tests
[16]	GWO-ANFIS controller	Develop maximum power point tracking (MPPT) for grid-tied EV-PV charging	Robust convergence to maximum power point (MPP) with better energy harvest	High implementation complexity; requires controller tuning

[17]	Metaheuristic optimization (Coati algorithm)	Mitigate EV load impact using optimal distributed generation (DG) allocation	Coati algorithm enhances voltage stability and power loss reduction	Results vary significantly with parameter settings
[18]	Empirical grid simulation	Assess ultra-fast-charging effects on grid pricing	Rapid EV charging distorts price signals and risks grid stability	Case-specific; may not apply to slower charging environments
[19]	Techno-economic simulation	Optimize PV-powered EV charging under climatic diversity	Regional solar profiles heavily affect station viability	Findings limited to Indian meteorological datasets
[20]	Predictive optimization [long short-term memory (LSTM)]	Integrate solar PV and EV in cold climate energy systems	LSTM improves prediction and operational efficiency	Cold-weather data needs further granularity for general adoption

Table 18.4 Consumer Trust in EV Technology by Age Group.

Age Group (Years)	*High Trust (%)*	*Moderate Trust (%)*	*Low Trust (%)*
18–34	62	29	9
35–54	49	37	14
55+	28	44	28

Range and charging concerns may deter rural or long-distance travelers. EV adoption depends on household structure, commute habits, and car usage if EV ownership is feasible.

Social Influence and Cultural Norms

Peer behavior, social media, and community adoption trends impact customer choices. Friends, family, and neighbors adopting electric cars raises the likelihood of other friends adopting them. In areas where municipal leaders mainstream or support EV adoption, consumers regard EVs as smart and innovative.

Table 18.5 data shows that social impact motivates new EV purchases. Meeting EV owners promotes confidence and reduces skepticism.

Charging Infrastructure Perception

Poor charging infrastructure, range anxiety, and charging infrastructure limitations hinder EV adoption. Drivers in flats or without private parking worry about charging locations and frequency. Public fast-charging stations, transparent pricing, and convenient parking locations can ease such concerns.

Psychological Ownership and Brand Attachment

Many car buyers consider brand reputation and personal attachment. While established automakers joining the EV business may exploit consumer trust, newer EV companies face skepticism. Psychological ownership or brand loyalty can affect EV adoption attitudes. Design, in-car tech, performance feedback, and brand story influence consumer attachment.

Table 18.6 illustrates that people choose traditional automakers for EVs because of their reputations, service networks, and brand familiarity. However, the EV market is growing rapidly, driven by both established manufacturers and inventive newcomers.

Psychological Barriers and Risk Perception

Uncertainty and risk hamper EV adoption. EV inexperience, technology obsolescence, battery longevity, range anxiety, and resale value are important psychological barriers. Many see transitioning from ICEVs to EVs as

Table 18.5 Effect of Social Influence on EV Purchase Likelihood.

Social Circle EV Ownership	*Purchase Likelihood (%)*
None	27
One or two individuals	51
Several individuals	68

Table 18.6 Brand Influence on Consumer EV Preference.

Brand Type	*Consumer Preference (%)*
Traditional auto brands (e.g., Ford, Volkswagen (VW)	54
New EV-only brands (e.g., Tesla, Rivian)	39
No brand preference	7

a financial and psychological risk. Risk aversion is reduced via warranties, customer education, satisfaction assurances, and performance data.

Conclusion and Synthesis

Economic, environmental, technological, psychological, and societal aspects affect electric car purchases. Buying decisions generally involve attitudes, expectations, experiences, and perceptions of the buyer. While early adopters have grown the EV market, widespread adoption requires addressing cost, infrastructure, trust, and lifestyle challenges. To boost electric mobility, manufacturers, legislators, marketers, and infrastructure planners must grasp these behavioral patterns. Consumer behavior can support sustainable transportation goals through targeted education, infrastructure improvements, economic incentives, and socially integrated promotions.

18.2 MARKET DRIVERS AND BARRIERS

Understanding EV market dynamics requires studying enabling drivers and persistent constraints. Government policy, technological innovation, infrastructure, economic conditions, and societal readiness shape these dynamics. This section uses data, comparative insights, and explanatory narratives to assess these market drivers and barriers, and empirical tables and extensive analysis to clarify and deepen their influences.

Government Incentives and Regulatory Support

Government policies shape EV ecosystems. Multiple state subsidies, tax exemptions, and restrictions have helped EV industry growth. Tax refunds, direct purchase subsidies, and lower registration fees reduce consumer expenses. Stringent emission regulations and sanctions for high-emission ICEVs also boost electrification. Government-funded charging infrastructure and research promote the sector.

Table 18.7 shows that countries that have successfully accelerated EV adoption utilize multifaceted policies, such as financial incentives, infrastructural investment, and regulatory support to boost EV markets.

Technological Advancements in Battery and Powertrain

Customer value has increased due to EV battery chemistry and electric drivetrain advances. Lithium-ion and solid-state batteries improve energy density, charging speed, and thermal management. These enhancements address consumer issues by improving vehicle range, charging speed, and performance. Over the past decade, battery costs have fallen, making EVs more accessible across markets.

From 2010 to 2024, battery pack prices fell from $1,100/kWh to $140/kWh. Lower costs narrow the EV–ICEV price gap. Modern lightweight materials, power electronics, and control algorithms make EVs more efficient and competitive.

Infrastructure Readiness and Urban Integration

EV adoption needs charging stations. In well-integrated cities, dense charging station networks simplify daily use and minimize range anxiety. Rural and underdeveloped areas have substantial charging infrastructure coverage gaps, discouraging buyers.

Table 18.8 shows the need for infrastructure development beyond metropolitan areas. Even when urban deployment drives EV adoption, long-term scalability requires equitable nationwide spread.

Table 18.7 Sample of Government Incentives for EVs in Leading Markets (2024).

Country	*Purchase Subsidy (USD)*	*Tax Rebate (%)*	*Emission Penalty for ICEVs*	*Charging Infrastructure Investment (USD Million)*
United States	7,500	10	Yes	4,500
Germany	9,000	19	Yes	3,200
China	4,000	13	Yes	6,000
Norway	0 (no VAT instead)	25	Yes	800

Table 18.8 Public Charging Station Density in Key Regions (Stations per 100,000 People).

Country	*Urban Areas*	*Rural Areas*	*Total National Average*
Netherlands	58	19	38
South Korea	42	15	31
United States	34	11	23
India	12	4	8

Market Competition and Innovation

New EV market entrants have increased competition and product innovation. Technology diversification by legacy automakers and EV start-ups offers vehicles for different customer demands, such as affordable compact automobiles, high-performance electric sedans, and commercial electric trucks. Consumers gain more choices, better features, and reduced prices.

Table 18.9 shows that R&D and market competition have continuously increased model availability. This growth represents early-stage innovation reaching mass commercial maturity.

Economic Volatility and Fuel Price Fluctuation

Economic conditions and fuel prices substantially impact consumer behavior. EV become more enticing when fuel prices are high and due to their lower operational costs. Electricity, increasingly generated from renewables, is more stable than gasoline or diesel. EV sales can reduce during recessions due to decreasing consumer purchasing power and vehicle financing.

As economic dynamics accelerate and hinder progress, EV adoption requires policy design. Zero-interest loans, leasing, and battery-as-a-service keep markets going during economic uncertainty.

Barriers: Range Anxiety and Charging Time

Range concern hinders EV adoption despite progress. Buyer expectations depend on ICEV refueling convenience, although many EVs cover over 300 kilometers per charge. Many consumers complain about Level 2 station recharging time.

Table 18.10 indicates that practical and psychological range and refilling convenience limitations still contribute to consumer reluctance, highlighting the need for infrastructural and battery improvements.

Table 18.9 Number of EV Models Available Globally by Year.

Year	*Number of Available EV Models*	*Major Contributors*
2015	37	Nissan, Tesla, BMW
2018	61	Hyundai, Kia, Chevrolet
2021	143	VW, Audi, Volvo, BYD
2024	234	Rivian, Lucid, XPeng, Ford, Mercedes-Benz

Table 18.10 Consumer Concerns Ranked by Severity (Survey of 2,000 Potential Buyers).

Concern	*Percentage of Respondents Ranking as Top Concern*
Range anxiety	38%
Charging time	26%
Vehicle price	19%
Battery lifespan and warranty	10%
Resale value uncertainty	7%

Table 18.11 Global Reserves and Market Share of Key Battery Materials (2024 Estimates).

Material	*Largest Reserve Holder*	*Estimated Global Reserve (Metric Tons)*	*Share of Supply (%)*
Lithium	Chile	9,200,000	43
Cobalt	Congo (DRC)	3,500,000	58
Nickel	Indonesia	21,000,000	34

Barriers: Supply Chain and Material Dependency

EV output is limited by supply shortages of lithium, cobalt, and nickel. This geographical concentration exposes the supply chain to geopolitical threats, pricing manipulation, and extraction-related environmental challenges. Decentralization, recycling, and lithium iron phosphate development mitigate the environmental impacts of the battery industry.

The data in Table 18.11 shows the key resource concentration, highlighting the need for material innovation and sustainable mining to sustain the business.

Market Sentiment and Media Influence

Media exaggerates EV achievements and failures, swaying consumers. Consumers notice technological improvements, celebrity endorsements, and policy achievements. Negative news like battery fires, charging troubles, or laws can slow EV adoption.

Table 18.12 Summary of Existing Analytical Models for Studying Market Dynamics.

Reference	*Method*	*Main Objectives*	*Findings*	*Limitations*
[21]	Survey and structural equation modeling	To examine how positive word-of-mouth (WOM) can paradoxically lead to switching intentions in mobile payments	Found a "boomerang effect" where overly positive WOM increases expectations, causing dissonance and switching	Limited to mobile payments, may not be generalizable to other sectors
[22]	*K*-nearest neighbors and ensemble stack generalization	To improve prediction accuracy for EV charging needs	Ensemble methods improved forecasting accuracy over standalone models	Computational complexity and real-time scalability issues
[23]	Game-theoretic modeling	To study green R&D strategies under asymmetric consumer preferences	Manufacturers prioritize green innovation when facing greener customer segments	Assumes rational behavior, not always realistic
[24]	Partial least squares structural equations modeling (PLS-SEM) and artificial neural network hybrid model	To analyze factors improving EV entrepreneurship dynamics	Technology orientation and government policy had the strongest positive effects	Data from emerging markets only
[25]	Hedonic pricing model	To assess the impact of EV charging station proximity on housing prices	Proximity to public EV stations increased property values in Beijing	Context-specific to urban China
[26]	Stochastic planning with carbon awareness	To integrate EV charging into renewable resource-based smart grids	Carbon-aware planning improves grid efficiency and emission targets	Model complexity limits practical deployment

(Continued)

Table 18.12 (Continued)

Reference	*Method*	*Main Objectives*	*Findings*	*Limitations*
[27]	Analytical calculations	To compute optimal motor rating and battery pack for EV-SUV	Proposed ratings optimize power efficiency and cost	Limited to a single vehicle segment
[28]	Market trend analysis	To review Russia's passenger car market and forecast EV trends	EV market remains nascent but shows policy-driven growth potential	Data availability and political instability concerns
[29]	Policy analysis	To design market-oriented electricity reforms	Incentive-based approaches foster renewable adoption	Does not directly address consumer EV behavior
[30]	System dynamics modeling	To simulate EV battery recycling outcomes under policy scenarios	Policy enforcement and subsidies enhance recycling rates	Model based on assumptions needing empirical validation
[31]	NBO-THDCNN algorithm	To manage energy of slow- and fast-charging EVs in smart parking	Achieved load balancing and improved charge control efficiency	Requires high computational infrastructure
[32]	Artificial intelligence (AI) with blockchain framework	To optimize demand response in EV charging networks	Enhanced security and real-time load balancing achieved	Blockchain scalability and interoperability challenges
[33]	Policy-market mechanism mapping	To explore private–public roles in energy transition	Effective partnerships crucial for EV infrastructure growth	Focuses more on theory than empirical validation
[34]	Techno-economic-environmental optimization	To distribute EV energy within smart grid using the innovative energy distribution method for electric vehicles (IEDM-EV)	Method offers improved grid stability and cost reduction	Requires centralized data and stakeholder cooperation

[35]	Hybrid simulation approach	To assess EV system with solar PV-integrated switched reluctance motor	Integration increased energy efficiency and reduced dependency	Hardware prototyping not included
[36]	Behavior-based vehicle-to-grid (V2G) simulation	To model power system operations with EV user behaviors	User behavior integration enhances grid reliability	Data collection on user habits is difficult
[37]	GJO-APCNN optimization	To improve EV charging station and grid performance	Optimized location and timing of charging improved grid utilization	Model testing in real-world settings pending
[38]	Neural network forecasting	To predict stock market using sentiment and technical indicators	Sentiment data improves financial forecast accuracy	Focused on finance, limited relevance to EV adoption
[39]	Dispatchable energy strategy simulation	To integrate EVs and storage into grid redundancy planning	Strategy reduces grid stress and improves energy reliability	Requires synchronized multi-source data streams
[40]	Moderated mediation modeling	To explore how sustainability practices affect market value	Revenue mediates and profitability moderates the effect of sustainability	Focus on firm-level metrics, not consumer behavior

Marketing must stress reliability, sustainability, and economic benefits to manage public opinion as seen in Table 18.12. Open communication, engaging content, and favorable testimonials promote confidence.

Summary and Strategic Implications

There are several drivers and challenges facing the electric car market. Positive momentum is fueled by increased competition, growing consumer awareness, technological improvements, and legislative support. However, expansion is still constrained by problems with material dependency, inadequate infrastructure, psychological barriers, and economic unpredictability. To solve these issues, a number of stakeholders—including governments, electricity providers, researchers, and automakers—must collaborate. Strong infrastructure, robust supply chains, and adaptable consumer education programs are essential to overcoming challenges and achieving the full potential of electric transportation. The strategic convergence of market forces, technology, and policy remains the cornerstone of an inclusive and sustainable EV transition.

REFERENCES

1. Shayeghi, H., & Davoudkhani, I. F. (2025). Uncertainty aware energy management in microgrids with integrated electric bicycle charging stations and green certificate market. *Scientific Reports*, 15(1). https://doi.org/10.1038/s41598-025-12328-5
2. Das, P., Gope, S., Das, D., Chakraborty, R., & Das, P. (2025). Optimal allocation of EV charging stations in a PV and wind energy sources integrated distributed system. *Electrical Engineering*, 107(9), 12669–12683. https://doi.org/10.1007/s00202-025-03175-8
3. Sudhakar, A., & Kumar, B. M. (2024). EV fleet energy management strategy for smart microgrids considering multiple objectives: Techno-economic perspective. *Arabian Journal for Science and Engineering*, 49(12), 16919–16939. https://doi.org/10.1007/s13369-024-09209-w
4. Jain, V. K., Kumari, A., Singh, S., & Tyagi, V. (2025). Electric vehicles as a pathway to decarbonisation in India: Assessing the environmental benefits and challenges of EV adoption for sustainable tomorrow. *Circular Economy and Sustainability*. https://doi.org/10.1007/s43615-025-00619-y
5. Bhattacharyya, S., Maurya, N., Jain, N., & Vishwakarma, D. K. (2024). Enhancing EV battery cooling using magnetic nanofluid and external magnetic field synergies. *Journal of Thermal Analysis and Calorimetry*, 149(24), 14971–14990. https://doi.org/10.1007/s10973-024-13829-7
6. Hao, X., Liu, P., Pu, H., Gong, F., Tong, F., Chen, Q., Liu, L., & Chen, X. (2025). Mobility-aware EV charging and discharging management in V2B-PV systems: A reinforcement learning framework. *Carbon Neutrality*, 4(1). https://doi.org/10.1007/s43979-025-00142-x

7. Pakbin, H., Karimi, A., & Hassanzadeh, M. N. (2025). An optimized demand response framework for enhancing power system reliability under wind power and EV-induced uncertainty. *Scientific Reports*, 15(1). https://doi.org/10.1038/s41598-025-05482-3
8. Bharti, A., & Chauhan, R. K. (2025). Optimized stochastic scheduling of wind energy with EV integration and demand response using MINLP. *Smart Grids and Sustainable Energy*, 10(3). https://doi.org/10.1007/s40866-025-00301-0
9. Katoch, R., & Peer, U. A. (2025). Navigating market risks in green investments in India: An evaluation of interest rate, equity, commodity, and Forex market influences. *Computational Economics*. https://doi.org/10.1007/s10614-025-11009-9
10. Demir, U., Ehsani, M., Demir, P., & Akpolat, A. N. (2025). Comparative analysis of ideal energy requirements in EV powertrains. *Iranian Journal of Science and Technology, Transactions of Electrical Engineering*, 49(3), 1337–1350. https://doi.org/10.1007/s40998-025-00809-2
11. Chhetri, A., Saini, D. K., Yadav, M., & Pal, N. (2024). Performance analysis of machine learning algorithms for estimation of EV penetration. *Microsystem Technologies*, 31(7), 1817–1831. https://doi.org/10.1007/s00542-024-05804-x
12. Shbool, M. A., Alhmsi, K., Amr, M., Hajeer, R., Hamed, A., Al-Bazi, A., & Rawabdeh, M. (2025). Modeling consumer behavior and forecasting the automobile market: A system dynamics approach for sustainable mobility. *Arabian Journal for Science and Engineering*. https://doi.org/10.1007/s13369-025-09984-0
13. Huang, X., Li, Z., Zhou, S., Liu, J., & Zhuge, C. (2025). Market potential of hydrogen fuel cell vehicles in Beijing: A spatial agent-based model approach. *Transportation*. https://doi.org/10.1007/s11116-025-10683-w
14. Dalkic-Melek, G., Saltik, E. C., & Tuydes-Yaman, H. (2024). Electric vehicle (EV) market penetration in countries with rising motorization rates. *International Journal of Civil Engineering*, 23(3), 461–480. https://doi.org/10.1007/s40999-024-01039-z
15. Rajakumari, R. F., Nithin, K., Jaganathan, S., Raja, M. A., & Devi, M. R. (2025). Performance analysis of modern EV batteries: Efficiency, longevity, and future innovations. *International Journal of System Assurance Engineering and Management*. https://doi.org/10.1007/s13198-025-02989-w
16. Mazumdar, D., Biswas, P. K., Sain, C., Ahmad, F., & Al-Fagih, L. (2024). A robust MPPT framework based on GWO-ANFIS controller for grid-tied EV charging stations. *Scientific Reports*, 14(1). https://doi.org/10.1038/s41598-024-81937-3
17. Painuli, S., Bhowmick, S., & Saha, R. (2024). Mitigation of residential EV charging effects on power distribution networks with optimal allocation of DGs using Coati optimization. *Arabian Journal for Science and Engineering*, 50(14), 10731–10750. https://doi.org/10.1007/s13369-024-09782-0
18. Yu, Q., Zhao, P., Li, J., Wang, H., Yan, J., & Zhang, H. (2025). China's urban EV ultra-fast charging distorts regulated price signals and elevates risk to grid stability. *Nature Communications*, 16(1). https://doi.org/10.1038/s41467-025-63199-3
19. Karmakar, A., Sadhu, P. K., Das, S., Bihari, S. P., Khan, B., & Ali, A. (2025). Techno-economic analysis and optimized PV-powered EV charging facilities under various climate conditions in India. *Discover Sustainability*, 6(1). https://doi.org/10.1007/s43621-025-01119-7

20. Hai, T., Ali, A. B. M., Agarwal, D., Punia, A., Jagga, M., Anqi, A. E., Ahmedi, M., Rajab, H., Singh, N. S. S., & Taghavi, M. (2025). Predictive optimization using long short-term memory for solar PV and EV integration in relatively cold climate energy systems with a regional case study. *Scientific Reports*, 15(1). https://doi.org/10.1038/s41598-025-01519-9
21. Pierre, S. J., & Mombeuil, C. (2025). The boomerang effect of positive word-of-mouth: Understanding switching intentions in the mobile payment market. *Journal of Financial Services Marketing*, 30(2). https://doi.org/10.1057/s41264-025-00308-9
22. Kumar, B. A., Jyothi, B., Singh, A. R., & Bajaj, M. (2024). Enhancing EV charging predictions: A comprehensive analysis using K-nearest neighbours and ensemble stack generalization. *Multiscale and Multidisciplinary Modeling, Experiments and Design*, 7(4), 4011–4037. https://doi.org/10.1007/s41939-024-00457-9
23. Wang, K., & Zhao, Y. (2025). Green technology research and development strategy choices of manufacturers under asymmetric consumer market. *Environment, Development and Sustainability*. https://doi.org/10.1007/s10668-025-06170-y
24. Prakhar, P., Jaiswal, R., Gupta, S., Ahmad, S. Z., Piccardi, P., & Santoro, G. (2025). How can we improve entrepreneurial dynamics in electric vehicle manufacturing for a sustainable future: Insights using a deep learning-based hybrid PLS-SEM-ANN approach. *International Entrepreneurship and Management Journal*, 21(1). https://doi.org/10.1007/s11365-025-01072-x
25. Zhao, X. (2025). Public charging stations and housing market: Evidence from Beijing. *Journal of Housing and the Built Environment*. https://doi.org/10.1007/s10901-025-10207-z
26. Wang, S., Tan, Q., Jiang, Q., Shao, H., & Kouhi, M. (2025). Stochastic carbon-aware planning of renewable DGs and EV charging stations with demand flexibility in smart urban grids. *Scientific Reports*, 15(1). https://doi.org/10.1038/s41598-025-17504-1
27. Jha, B. K., & Bohre, A. K. (2024). Motor rating and battery pack calculation for an EV-SUV. *SN Computer Science*, 5(1). https://doi.org/10.1007/s42979-023-02552-z
28. Milyakin, S. R., Skubachevskaya, N. D., & Migal', A. V. (2025). The passenger car market in Russia: History, current status and forecast. *Studies on Russian Economic Development*, 36(1), 86–95. https://doi.org/10.1134/s1075700724700552
29. Li, Q., & Xi, W. (2024). Sustainable and market-oriented solutions for the electricity sales market reform. *Economic Change and Restructuring*, 57(2). https://doi.org/10.1007/s10644-024-09608-x
30. Wei, L. (2025). How to facilitate the recycling of retired electric vehicle batteries in China? A policy simulation based on system dynamics. *Journal of Material Cycles and Waste Management*, 27(5), 3749–3761. https://doi.org/10.1007/s10163-025-02330-4
31. Devi, B., & Jose, S. E. (2024). Energy management of slow and fast charge electric vehicles (EV) in smart parking lot using NBO-THDCNN algorithm. *Environment, Development and Sustainability*, 27(8), 20047–20070. https://doi.org/10.1007/s10668-024-05241-w

32. Singh, A. R., Kumar, R. S., Madhavi, K. R., Alsaif, F., Bajaj, M., & Zaitsev, I. (2024). Optimizing demand response and load balancing in smart EV charging networks using AI integrated blockchain framework. *Scientific Reports*, 14(1). https://doi.org/10.1038/s41598-024-82257-2
33. Song, T., Li, H., & Feng, Z. (2024). Policy and market mechanisms for promoting sustainable energy transition: Role of government and private sector. *Economic Change and Restructuring*, 57(4). https://doi.org/10.1007/s10644-024-09734-6
34. Jagan, C., Sivaraju, S. S., Thangavelu, A., & Samidurai, S. (2025). Innovative energy distribution method for electric vehicles (IEDM-EV): A multi-objective techno-economic-environmental optimization for smart grid integration. *The International Journal of Advanced Manufacturing Technology*. https://doi.org/10.1007/s00170-025-15359-4
35. Devi, B., & Edwin Jose, S. (2025). Efficiency assessment of EV system integrated with solar PV based SRM using hybrid approach. *Environment, Development and Sustainability*, 27(10), 25679–25698. https://doi.org/10.1007/s10668-025-05977-z
36. Shin, G., Kim, H., Mahseredjian, J., & Kim, C. (2024). Smart vehicle-to-grid operation of power system based on EV user behavior. *Journal of Electrical Engineering & Technology*, 19(5), 2941–2952. https://doi.org/10.1007/s42835-024-01797-9
37. Gunapriya, B., Kumar, B. S., Rajalakshmi, B., & Amarendra, A. (2024). Performance enhancement of EV charging stations and distribution system: A GJO–APCNN technique. *Electrical Engineering*, 107(1), 883–897. https://doi.org/10.1007/s00202-024-02531-4
38. Arauco Ballesteros, M. A., & Martínez Miranda, E. A. (2024). Stock market forecasting using a neural network through fundamental indicators, technical indicators and market sentiment analysis. *Computational Economics*, 66(2), 1715–1745. https://doi.org/10.1007/s10614-024-10711-4
39. Ajao, Q., & Sadeeq, L. (2025). A rapidly dispatchable energy strategy utilizing electric vehicles with energy storage systems (EV-ESS) to enhance grid redundancy. *Smart Grids and Sustainable Energy*, 10(1). https://doi.org/10.1007/s40866-025-00243-7
40. Komath, M. A. C., Sayilir, Ö., & Doğan, M. (2025). The relationships between corporate sustainability and market value: Mediating role of revenue and moderating role of profitability. *International Journal of Disclosure and Governance*. https://doi.org/10.1057/s41310-025-00288-1

Chapter 19

Conclusion and Future Outlook

19.1 KEY TAKEAWAYS

Introduction: Synthesizing the Landscape of Electric Mobility

This comprehensive book on electric vehicles (EVs) concludes with a summary of the key findings, discoveries, and implications. Market factors, technology developments, legislative changes, user behavior, and environmental impacts make the transfer to electric mobility from internal combustion engine vehicles (ICEVs) difficult. Researching EV technologies, case studies, and application frameworks can help one understand the automobile industry's future throughout this paradigm shift. To demonstrate the book's depth, this chapter discusses the significant breakthroughs, challenges, and future of electric mobility.

Historical Evolution and Foundational Principles

EV history indicates that electric drivetrains are 19th-century ideas. Technological limits and gasoline supply prohibited electric alternatives for over a century. The context has changed due to battery energy density, power electronics, and global environmental awareness. EVs use electromotive force, regenerative braking, and energy storage. Chapter One clearly explained how technical preparedness, market conditions, and policy frameworks enable industrial revolution.

Table 19.1 presents this transition using historical milestones.

Table 19.1 shows that EV success depends more on technology maturity and systems integration than novelty. EVs have only recently created a mass-use environment due to battery cost decrease, legislative support, and expanded charging infrastructure.

Technical Architecture and Component Advancements

The EV architecture is modular and efficient. The chapters on technical architecture focus on battery, electric motor, inverter, and power electronics

 DOI: 10.1201/9781003608097-19

Table 19.1 Timeline of EV Development and Milestones.

Year	*Milestone Description*
1830s	First small-scale EVs built using non-rechargeable cells
1889–1900	Commercial electric carriages gain popularity in urban settings
1912	Charles Kettering invents the electric starter, boosting ICEV practicality
1970s	Oil crises revive interest in alternative propulsion systems including electric cars
1997	Launch of Toyota Prius, marking mainstream acceptance of hybrid technology
2010	Nissan Leaf becomes one of the first mass-produced battery electric vehicles (BEVs)
2012	Tesla Model S reshapes public perception and expectations for EV performance
2020s	Major automakers announce plans to phase out ICEVs in favor of EVs

synergy. These components have been modified independently to improve system performance, range, and reliability.

Three-phase alternating current (AC) induction and permanent magnet synchronous motors (PMSM) have replaced brushed direct current (DC) in EV motors. This improved torque, motor efficiency, and thermal control. Battery packs are progressing from lead-acid to advanced lithium-ion chemistries to safe, high-energy solid-state batteries. Table 19.2 compares EV components by generation, demonstrating core technical advancement.

Table 19.2 indicates that EV architectural progress is diverse. Generational improvements increase performance, safety, scalability, and cost. These trends aid stakeholders in technology selection and strategy futureproofing.

Battery Systems and Charging Infrastructure

The book emphasizes that battery performance and charging infrastructure promote EV adoption. Battery technology both drives and holds back the EV industry. Battery energy density, charging time, lifetime, safety, and cost per kilowatt hour determine EV usefulness and economic feasibility.

Charging infrastructure links EV technology and trust. The density, accessibility, and compatibility of public and private charging networks affect user choices and range anxiety. Table 19.3 shows that battery characteristics and charging infrastructure affect consumer experience.

Table 19.3 indicates that battery and charging technologies must be developed jointly. Maximizing electric mobility requires a powerful battery and scalable charging infrastructure. Because of these interdependencies, hardware and grid-level design and policy must be integrated.

Table 19.2 Comparison of Key EV Components by Generation.

Component	*Early Generation (Pre-2010)*	*Current Generation (2020s)*	*Emerging/Future Generation*
Battery type	Lead-acid, nickel–metal hydride (NiMH)	Lithium ion (lithium nickel manganese cobalt oxide (NMC), lithium iron phosphate (LFP))	Solid-state, lithium–sulfur
Motor type	Brushed DC	AC induction, PMSM	Axial flux, in-wheel motors
Inverter tech	Basic insulated gate bipolar transistors (IGBT)	High-efficiency SiC metal-oxide-semiconductor field-effect transistors (MOSFETs)	Gallium nitride (GaN)-based high-frequency inverters
Charging capability	Level 1 (slow)	Level 2 and DC fast charging	Ultra-fast, wireless charging
Battery management system (BMS) features	Basic monitoring	Advanced state of charge (SoC), state of health (SoH), thermal control	Artificial intelligence (AI)-enhanced predictive BMS

Table 19.3 Interplay Between Battery and Charging System Characteristics.

Parameter	*Battery Specification*	*Charging System Requirement*	*User Impact*
Energy density	Higher (Wh/kg)	Support for faster charge rates	Increased range, reduced charge frequency
Thermal stability	Improved cell design	Real-time thermal control during charging	Enhanced safety, especially during fast charging
Charging time	Rapid charging chemistries	High-power (direct current fast chargers (DCFCs)) infrastructure	Shorter downtime, improved convenience
Cycle life	Durable chemistries (e.g., LFP)	Adaptive charging profiles	Lower total cost of ownership
Degradation rate	Low-capacity fade batteries	Smart charging (avoid overcharging)	Longer lifespan, better resale value

Environmental and Economic Implications

The environmental impact of electric vehicles goes beyond emissions. The book's lifecycle assessments (LCAs) indicate that EVs' true ecological benefit depends on the electrical grid's carbon intensity, battery manufacture, and end-of-life recycling. When optimized, EVs exceed ICEVs in greenhouse gas (GHG) emissions and energy efficiency.

EVs are cheaper to operate because to fewer moving parts, little maintenance, and lowering electricity prices. Policy involvement is needed in many markets with high starting costs. Table 19.4 compares the lifecycle environmental impacts of ICEVs and EVs.

Table 19.4 shows that EVs' environmental benefits increase over time, especially when powered by low-carbon electricity. End-of-life innovation is needed, notably in battery recycling and circular economy models.

Market Penetration, Policy Support, and Consumer Education

The book argues that technological readiness does not guarantee market success. Subsidies, regulations, and public awareness strongly influence EV adoption. Without subsidies, tax incentives, and urban automobile bans, EV adoption would stagnate. Demystifying EV ownership through consumer education strengthens the impact of these measures.

Global market data reveals that policy-supported regions have more EVs. In Scandinavian countries with big incentives, comprehensive charging infrastructure, and public campaigns, EVs exceed 70% of new car sales, whereas those without policy engagement have single-digit figures.

Table 19.5 indicates that multidimensional regulatory frameworks, consumer awareness, and infrastructure preparation boost EV adoption. Countries that view EVs as a system-level revolution rather than a product-level substitution do better.

Table 19.4 Comparison of Lifecycle Environmental Impacts of ICEVs versus EVs.

Lifecycle Stage	*ICEV*	*EV*
Manufacturing emissions	Moderate	High (battery production intensive)
Operational emissions	High (tailpipe CO_2, NOx)	Low to zero (depends on grid mix)
Fuel source	Fossil fuels	Electricity (renewable or mixed grid)
Maintenance requirements	High (oil changes, engine wear)	Low (few moving parts, no oil system)
End-of-life impact	Standard recycling, moderate waste	Requires battery recycling infrastructure

Table 19.5 Comparison of EV Adoption Drivers and Outcomes by Region.

Region	*Policy Support Strength*	*Charging Infrastructure Density*	*Public Awareness*	*EV Market Share (2024 Est.)*
Norway	Very high	High	High	80%
China	High	Medium	Medium	35%
United States	Medium	Low to medium	Medium	12%
India	Low	Low	Low	2%
Germany	High	High	High	22%

Future Outlook and Transformative Potential

The EV industry will change. Vehicle-to-grid (V2G) communication, solid-state batteries, and electric drivetrain autonomous driving seem promising. A solid-state battery could alleviate liquid electrolyte safety and longevity issues. V2G technologies make EVs mobile energy assets, stabilizing networks and improving efficiency. Electric platforms may facilitate autonomous system integration due to their digital control systems and reduced mechanical complexity.

Electric mobility will transform transportation, not just replace ICEVs. Electric mobility impacts renewable energy, digital infrastructure, and urban planning. Future smart cities will use electric, shared, and autonomous mobility. This future-focused analysis reveals that EVs help decarbonization and sustainability as a transition technology.

This book promotes cross-disciplinary EV research. This sector needs technology, law, user involvement, and sustainability, not just batteries or motors. Auto, energy, electronics, and software industries converge in electric cars. The main findings suggest that electric mobility is feasible and necessary for climate resilience, urban health, and economic sustainability.

These findings reflect a bigger shift for varied conditions. They prepare for research, innovation, and policy. The lessons from this work reflect past accomplishments and influence future development.

19.2 ELECTRIC VEHICLES IN REDUCING GREENHOUSE GAS EMISSIONS

Introduction: The Climate Imperative and Role of EVs

GHG emissions are crucial to energy and transportation policy due to climate change. One-quarter of global CO_2 emissions originate from transportation, making it a major target for carbon footprint reduction. As an environmental and technological advance, EVs have garnered attention. Unlike ICEVs, EVs might eliminate fossil fuel burning in personal and business transportation. Energy supply, vehicle efficiency, manufacturing practices, and energy policy integration determine this potential. EVs reduce GHG emissions through lifecycle evaluation, energy grid dynamics, traffic patterns, and long-term emission modeling.

Lifecycle Greenhouse Gas Emissions: A Systemic Analysis

Understanding EV emissions requires LCA), which evaluates a product's environmental impact from raw material extraction to disposal. LCA says EVs have zero tailpipe emissions but are not emission-free. Battery manufacturing and charging electricity generation cause most EV emissions. EVs emit less over time than ICEVs, especially in low-carbon electricity regions.

Electric car production emits more CO_2 due to energy-intensive battery production (Table 19.6). EVs emit less CO_2 when charged from low-carbon or renewable grids. After 150,000 kilometers, EVs release 50–70% less GHGs than gasoline-powered vehicles. It appears that EVs have long-term environmental benefits despite higher initial emissions.

Grid Emissions Intensity: The Role of Electricity Mix

The environmental impact of EVs depends on the carbon intensity of the charging electricity. BEVs charged by coal-dominated power may pollute more than efficient hybrids. Hydropower, wind, and solar-charged EVs emit less. EVs' net GHG reduction potential varies by region, time, and energy policy.

The average grid emissions intensity (gCO_2e/kWh) differs greatly between countries and regions. This intensity affects EVs' effective emissions per kilometer. Regional grid mix EV emissions are compared in Table 19.7.

The table illustrates that EVs almost eliminate driving-related GHG emissions in Norway and France with clean grids. In coal-dependent India and

Table 19.6 Comparison of Lifecycle Emissions of ICEVs versus EVs (per Vehicle over 150,000 km).

Emission Category	*ICEV (Gasoline)*	*BEV (Grid Avg.)*	*BEV (Renewable Grid)*
Manufacturing emissions (kg CO_2e)	7,000	10,500	10,500
Operation (fuel/charging) (kg CO_2e)	33,000	11,000	2,000
Maintenance (kg CO_2e)	1,000	600	600
Total lifecycle emissions	41,000	22,100	13,100

Table 19.7 Comparison of Estimated Operational Emissions of EVs Based on Grid Carbon Intensity.

Region	*Grid Carbon Intensity (gCO_2e / kWh)*	*EV Emissions (gCO_2e / kWh)*	*EV versus ICEV Emissions (%)*
Norway	20	6	98% lower
France	60	18	94% lower
Germany	400	120	65% lower
United States	380	114	67% lower
India	720	216	25% lower
China	600	180	40% lower

China, EVs reduce pollution only to a small degree. EVs' environmental benefits increase with grid decarbonization.

Transportation Sector Decarbonization Scenarios

Beyond car assessments, macro-level modeling is needed to understand how EVs affect national and global GHG emissions. International climate models suggest large-scale EV adoption may be necessary to reach net-zero by mid-century. These achievements require concurrent supportive infrastructure, renewable energy deployment, and fleet electrification rules.

Decarbonization possibilities range from modest penetration to aggressive policy-led changes. EV sales will approach 70% in high-adoption models by 2040, replacing millions of ICEVs and reducing emissions by 60%. Table 19.8 shows three main climate research consortia's global scenarios.

Table 19.8 demonstrates that strong EV adoption requires large grid emission reductions. Most vehicles must be electric, and electricity must come from low-carbon sources such as wind, solar, hydro, or advanced nuclear. Transport electrification and power sector decarbonization must occur simultaneously to reduce GHGs.

Urban Air Quality and Non-CO_2 Emission Benefits

Climate policy prioritizes CO_2, but co-pollutants like NO_x, PM2.5, PM10, and volatile organic compounds (VOCs) from ICEVs significantly affect urban air quality. Non-burning EVs minimize pollution and improve public health in densely populated areas.

Oslo, Beijing, and San Francisco, which vigorously promote EV use, have improved air quality and reduced respiratory disorders and asthma-related hospital admissions. These non-GHG co-benefits of EV deployment promote climate goals and provide immediate socioeconomic benefits.

Table 19.8 Projected Global Emissions Reduction by 2050 under EV Adoption Scenarios.

Scenario Type	*EV Share of New Sales by 2040*	*Total Transportation Emissions Reduction*	*Required Grid Carbon Intensity*
Business-as-usual (BAU)	20%	15% reduction	~500 gCO_2e/kWh
Moderate policy scenario	50%	35% reduction	~300 gCO_2e/kWh
Net-zero target scenario	75–90%	60–70% reduction	~100 gCO_2e/kWh

Battery Recycling, Circular Economy, and Net Emissions

EV battery manufacturing, especially lithium, cobalt, and nickel mining, is an environmental concern. Advanced battery recycling and second-life applications are rapidly lowering these consequences. Sustainable electric mobility requires a circular battery economy that reuses end-of-life EV batteries for grid storage or recycles them for raw materials.

Direct cathode reuse and hydrometallurgical recovery reduce embedded emissions from battery recycling. Firms must also recycle according to EU and North American rules. Battery production is becoming less GHG-intensive with each generation.

Policy Instruments to Maximize Emissions Reduction

Technology and policy determine EVs' climate mitigation potential. Carbon price, purchase incentives, zero-emission vehicle legislation, fuel economy standards, and charging infrastructure investment affect emissions. These levers encourage customers to utilize vehicles more efficiently and sustainably.

Policies targeting ride-sharing fleets, public buses, and commercial delivery trucks reduce emissions more due to their higher use. Coordination among energy, transportation, and urban development is optimal for decarbonization.

Conclusion: Quantifying the Impact and Addressing the Limits

Overall, EVs' GHG emission reduction benefits are clear yet conditional. When grid electricity cleans up, EVs outperform ICEVs in lifecycle GHG emissions. Regional energy condition, vehicle class, user behavior, and policy environment affect this benefit. The tables in this section illustrate that EVs are crucial to climate policy but must be implemented within power sector decarbonization, sustainable materials procurement, and urban transportation reform.

Although not perfect, EVs are essential to reducing global pollution. When widely used in freight, transit, logistics, and personal travel, EVs reduce emissions most. Smart grid integration, renewable energy alignment, and lifecycle optimization increase these benefits. Thus, EVs' environmental justification is strong and expanding, boosting their sustainable mobility role.

19.3 TRANSFORMING THE AUTOMOTIVE INDUSTRY

Introduction: A Paradigm Shift in Motion

ICEV to EV transition is altering the global car industry in ways not seen in a century. This transformation affects industrial processes, supply chains, labor dynamics, market structures, regulatory frameworks, and customer expectations, not just propulsion technology. Electricity is accompanied by a digital revolution involving connected cars, over-the-air (OTA) updates, autonomous driving, and new ownership structures. All of these developments require a reconsideration of the automobile, its production, use, and value chain. EVs are changing the auto industry, altering manufacturing, employment, business models, global trade, and industrial strategy.

Manufacturing Transformation: From Engines to Electronics

Traditional ICE car manufacturing uses crankshafts, camshafts, pistons, gearboxes, and exhaust systems. These labor-intensive components are supplied by a wide, specialized supplier network. Electric cars have fewer moving parts and a simpler drivetrain. Electric motors, inverters, and battery packs replace complex mechanical components, reconfiguring production lines and sourcing.

Manufacturing EVs is more like electronics than vehicles. Giant factories with cleanrooms, chemical handling, and robotics make batteries and packs. Automotive and electronics manufacturing are converging, redefining industrial skills. Table 19.9 shows how ICE and EV component architectures differ, illustrating production process restructuring.

Simplifying EV component categories reduces part count by 30–40%, simplifying assembly and supply chains. Industrial organization suffers, notably in ICE component manufacturing locations and businesses.

Table 19.9 Comparison of Core Components in ICEV Versus EV Manufacturing.

Component Category	*ICEVs*	*EVs*
Powertrain	Engine block, pistons, valves, crankshaft	Electric motor, inverter, reduction gear
Transmission	Manual/automatic gearbox, torque converter	Single-speed transmission
Fuel/energy system	Fuel tank, fuel pump, injectors	Battery pack, BMS
Exhaust and emissions	Exhaust manifold, catalytic converter	Not required
Thermal management	Radiator, engine coolant loop	Battery and motor thermal management units
Control systems	Electronic control units (ECUs) (engine control), anti-lock braking system (ABS)	Microcontroller unit (MCU), integrated software stack

Supply Chain Realignment and Raw Material Dependencies

Electromobility has shifted supply chain focus from mechanical engineering to materials research. ICE cars needed steel, aluminum, and polymers. EVs require huge lithium, cobalt, nickel, and rare-earth elements for battery cathodes and permanent magnet motors. This transformation raises geopolitical and ethical concerns for the auto industry, particularly in mineral sourcing and labor.

Major automakers are vertically integrating battery supply chains through strategic partnerships to ensure long-term ingredient access. Gigafactory projects are near mining and refining areas to avoid price volatility and supply disruptions. Table 19.10 shows escalating key mineral requirements for EV production and associated sourcing implications.

Table 19.10 demonstrates that a few countries dominate mineral extraction and processing, many with fragile regulatory frameworks or poor labor practices. Technology and supply chain governance ethics and geopolitics must alter to transform the auto industry.

Workforce Implications and Labor Realignment

The switch from ICEV to EV production hurts the auto industry. Electric drivetrains are simpler, so assembly lines are less labor-intensive, threatening engine and transmission hub jobs. Battery cell manufacturing, power electronics, software engineering, and charging infrastructure development jobs result from this shift.

Table 19.10 Critical Minerals for EV Production and Global Supply Concentration.

Mineral	*Primary EV Application*	*Top Producer Countries (%)*	*Strategic Concerns*
Lithium	Battery cathodes (LFP, NMC)	Australia (53%), Chile (26%), China (13%)	High demand growth, water-intensive mining
Cobalt	Battery cathodes (NMC)	DR Congo (70%), Russia (5%), Australia (4%)	Ethical sourcing, child labor risks
Nickel	Battery cathodes (NCA, NMC)	Indonesia (34%), Philippines (11%), Russia (9%)	Emissions from laterite processing
Graphite	Battery anodes	China (65%), Mozambique (12%), Brazil (9%)	Supply chain concentration
Rare earths	Electric motors (NdFeB magnets)	China (86%), United States (10%), Myanmar (3%)	Export restrictions, recycling challenges

Just transition requires worker reskilling. Former thermodynamics and combustion engineers study electrical engineering, battery chemistry, and systems integration. EVs' embedded software—from car diagnostics to autonomous driving—has increased the need for digital capabilities in a mechanical-dominated industry. The EV-era skill transition matrix for workforce realignment is shown in Table 19.11.

Educational institutions, industry associations, and governments must coordinate upskilling to satisfy car production needs. If unchecked, this workforce movement could worsen structural unemployment and economic inequality in auto manufacturing locations.

New Business Models and the Digital Vehicle

Electric cars change car manufacturing, sales, ownership, and monetization. EVs' software-centric architecture allows direct-to-consumer sales, Vehicle as a Service (VaaS), subscription-based features, and OTA upgrades. Software-updated EVs can produce continual revenue, unlike ICEVs that need hardware upgrades for performance or features.

Automakers are increasingly embracing the car as a digital platform. Vehicle software offers infotainment, navigation, battery management, telematics, and predictive maintenance. Table 19.12 provides a comparison of electric and traditional mobility ecosystem revenue models.

Manufacturers must use robust encryption, authentication, and software validation for OTA functionality to meet cybersecurity and regulatory requirements. Automakers must compete with digital companies in user experience, data management, privacy, and platform ecosystems as hardware sales shifts to software.

Table 19.11 Skill Transition Matrix for Automotive Workforce Transformation.

Traditional Role (ICE Era)	*Emerging Role (EV Era)*	*Skill Gap Addressed*
Engine assembly technician	Battery pack technician	Electrical safety, lithium cell handling
Transmission design engineer	Electric motor control engineer	Power electronics, control algorithms
Exhaust system technician	Thermal management systems specialist	Heat pump systems, coolant design
Mechanical QA analyst	Software QA and HIL testing engineer	Python, MATLAB, simulation frameworks
Fuel systems engineer	BMS developer	Embedded C, SoC algorithms, thermal control

Table 19.12 Revenue Model Shift in the Automotive Industry.

Revenue Source Category	*ICEV Model*	*EV and Software-Defined Vehicle Model*
Initial sale	Primary revenue	Still significant, often direct-to-consumer
Aftermarket services	Repairs, parts replacement	Minimal due to low maintenance
Fuel/energy	External (gas stations)	Integrated (charging networks, apps)
Upgrades and features	Hardware-based, physical installs	Software-based, OTA updates, subscriptions
Data monetization	Minimal	Increasingly central (usage patterns, Advanced driver-assistance systems (ADAS))

Geopolitical Repositioning and Industrial Strategy

Global industry and national strategy are shifting with electric transportation. Countries use industrial strategy to ensure battery, semiconductor, and mineral access, not just car exports. European, US, Chinese, South Korean, and Japanese industrial strategies include subsidies, R&D incentives, and trade barriers to lead EVs.

China, which manufactures EVs from mining to assembly, is pushing Germany, the United States, and Japan for automotive leadership. Southeast Asian countries are becoming battery material refiners and auto assembly hubs. Reestablishing industrial leadership in the EV era requires long-term investment, strategic autonomy in materials, and technological sovereignty.

Conclusion: A Rewiring of the Automotive DNA

Electric cars are changing the auto industry. Manufacturing, supply chain, labor, and revenue generation are all being restructured in the automobile value chain. Sustainability, digitization, and electrification drive this transition. EVs accelerate this shift and have unprecedented industrial, consumer, and geopolitical capabilities.

The auto industry currently operates in a mobility ecosystem with energy, digital, and environmental regulations. This structural alteration illustrates that cross-domain adaptation, not mechanical engineering, drives EV success. Automotive leadership will transition to battery, software, ethical sourcing, and digital service firms.

Appendices

Appendices

A. Glossary of Terms

Advanced Driver-Assistance Systems (ADAS): Smart technologies, such as data analytics, designed to improve vehicle safety and functionality through real-time data (p. 10).
Battery Electric Vehicle (BEV): A vehicle exclusively powered by rechargeable battery packs, featuring zero tailpipe emissions (pp. 15, 28).
Battery Management System (BMS): A specialized system used to monitor and manage the performance, safety, and longevity of an EV's battery pack (p. 7).
Fuel Cell Electric Vehicle (FCEV): A vehicle that uses hydrogen fuel cells to generate electricity for propulsion, emitting only water vapor as a byproduct (pp. 28, 37).
Hybrid Electric Vehicle (HEV): A vehicle that integrates both an internal combustion engine and an electric motor; it cannot be charged externally and relies on regenerative braking (pp. 16, 28).
Plug-in Hybrid Electric Vehicle (PHEV): A vehicle combining a traditional engine with an electric motor and a battery that can be recharged from an external source (pp. 15, 31).
Power Electronics: Components that control and regulate the flow of electrical energy between the battery and the electric motor (p. 20).
Regenerative Braking: A mechanism that recaptures kinetic energy during deceleration and converts it back into electrical energy to recharge the battery (pp. 21, 44).
Solid-State Battery (SSB): An emerging battery technology using solid electrolytes, offering higher energy density and improved safety over traditional liquid-electrolyte batteries (p. 10).
Vehicle-to-Grid (V2G): Technology allowing electric vehicles to communicate with and provide energy back to the power grid (p. 3).

B. List of Abbreviations

ADAS	Advanced Driver-Assistance Systems (p. 5)
BEV	Battery Electric Vehicle (p. 2)
BMS	Battery Management System (p. 7)
FCEV	Fuel Cell Electric Vehicle (p. 2)
HEV	Hybrid Electric Vehicle (p. 2)
ICE	Internal Combustion Engine (p. 3)
NOx	Nitrogen Oxides (p. 22)
PHEV	Plug-in Hybrid Electric Vehicle (p. 2)
PM	Particulate Matter (p. 22)
SSB	Solid-State Battery (p. 10)
V2G	Vehicle-to-Grid (p. 3)
VOCs	Volatile Organic Compounds (p. 22)

C. Additional Resources

International Energy Agency (IEA): For global market data, deployment trends, and policy forecasts (p. 8).

ISO & SAE Standards: Specifically, ISO 26262 for safety and SAE J1772 for charging connectors (p. 9).

Greenhouse Gas (GHG) Calculators: Tools like the GREET model for conducting Life Cycle Assessments (LCAs) of EV emissions (pp. 7, 21).

Consumer Support: Charging station databases (e.g., Plug Share) and total cost of ownership (TCO) calculators (pp. 9–10).

Index

advanced driver-assistance systems (ADAS) 40
aerodynamics 77
affordability and accessibility 181
artificial intelligence (AI) in EVs 134, 254
autonomous driving 134

battery electric vehicles (BEVs) 3, 14, 119
battery management systems (BMS) 44, 55
battery packs: chemistry 30, 44; degradation 29, 90; recycling and sustainability 177; thermal management 44

carbon emissions 8, 105
charging infrastructure: fast-charging networks 69, 131; home charging 63; public charging 66; standards and compatibility 72, 209; wireless charging 245
consumer behavior 273
consumer education 208
control systems 41

drivetrains 26, 40

electric motors 2, 26, 40
energy management systems 84
environmental impact 8, 105, 111

fuel cell electric vehicles (FCEV) 21, 28

government policies and incentives 120, 123
greenhouse gas (GHG) emissions 8, 294

historical evolution (EV) 1, 4, 18
hybrid electric vehicles (HEV) 3, 19, 119

internal combustion engine (ICE) comparison 101, 111

lifecycle assessment 108
lithium-ion batteries 19, 46

maintenance (EV) 90
market trends 119

plug-in hybrid electric vehicles (PHEV) 3, 17
power electronics 34, 41
propulsion systems 23, 27

range anxiety 217
regenerative braking systems 37, 81
renewable energy sources 105

safety standards 189, 202
solid-state batteries 51, 131, 240

vehicle-to-grid (V2G) integration 136

For Product Safety Concerns and Information please contact our
EU representative GPSR@taylorandfrancis.com Taylor & Francis
Verlag GmbH, Kaufingerstraße 24, 80331 München, Germany